Organized Technology

Networks and Innovation in Technical Systems

by Wesley Shrum

Purdue University Press
West Lafayette, Indiana
1985

Library of Congress Cataloging in Publication Data

Shrum, Wesley, 1953-
Organized technology.

Bibliography: p.
Includes index.
1. Technological innovations--United States.
2. Photovoltaic power generation--United States
3. Radioactive waste disposal--United States.
I. Title.

T173.8.S54 1985 306'.46 85-3553
ISBN 0-911198-74-1

Printed in the United States of America

To Wesley Monroe Shrum, Kathryn, and Kem

Contents

Exhibits

Acknowledgments

Many were responsible for what follows. Our informants, participants in the development of nuclear waste and solar cell technology, deserve the first thanks, or at least 93% of them do. "Our," throughout the book, refers to the project team of Robert Wuthnow, James Beniger, Patricia Woolf, Laura Schrager, Cathy Leeco, and Karen Cerulo. Together, they made the Technical Systems Project a success.

Of course, modern quantitative social science is no longer a shoestring operation. In our case funds were provided by the National Science Foundation (PRA-7920573; Robert Wuthnow, Principal Investigator). The Innovation Processes Section, consisting of Louis Tornatzky, J.D. Eveland, William Hetzner, and Myles Boylan, was probably as fine a group as you could ask for, bar none.

There are two people who merit special thanks. The first is Robert Wuthnow, whose intellectual flexibility and assistance in the present work were boundless. Our understanding of modern culture would be greatly enhanced if more sociologists explored, as he does, the nature of both science and religion. The second is James Beniger, who taught me respect for the things numbers can teach us and disrespect for received wisdom.

A number of people read various parts of the manuscript and provided useful comments, including Michael White, Marvin Bressler, David Crerar, David Redfield, Sigurd Wagner, William Bankston, Forrest Deseran, Thomas Durant, William Falk, Michael Grimes, Lisandro Perez, and, in a different version, Mark Fossett, Jill Kiecolt, and Linda Stearns. Reba Rosenbach produced a quality manuscript for the final version of the book. Lee Trachtman and Verna Emery at Purdue University Press presided over the review and production of the book, respectively.

Finally, I would like to thank John Kuzloski, Kevin Christiano, and Robert Cox, whose cavalier disregard for rational procedure hastened completion of the work.

Varna, Bulgaria
September, 1984

Acronyms

A

ACE	Army Corps of Engineers
AEC	Atomic Energy Commission
AFR	Away From Reactor
APS	American Physical Society

B

BES	Office of Basic Energy Sciences
BWIP	Basalt Waste Isolation Project

D

DOD	Department of Defense
DOE	Department of Energy
DOT	Department of Transportation

E

EOP	Executive Office of the President
EPA	Environmental Protection Agency
ERDA	Energy Research and Development Administration

F

FY	Fiscal Year

H

HLW	High-Level Wastes

I

IRG	Interagency Review Group on Nuclear Waste Management

J

JPL	Jet Propulsion Laboratory

L

LLW	Low-Level Wastes

N

NAS	National Academy of Sciences
NASA	National Aeronautics and Space Administration
NNWSI	Nevada Nuclear Waste Storage Investigations
NRC	Nuclear Regulatory Commission
NSF	National Science Foundation
NWTS	National Waste Terminal Storage

O

OMB	Office of Management and Budget
ONWI	Office of Nuclear Waste Isolation
ONWM	Office of Nuclear Waste Management
OTA	Office of Technology Assessment

P

PERT	Program Evaluation and Review Technique

R

R&D	Research and Development
RFP	Request For Proposals
RSSF	Retrievable Surface Storage Facility

S

SERI	Solar Energy Research Institute

T

TRU	Transuranic (waste)

U

USGS	U.S. Geological Survey

Introduction

Chapter 1

Technology and Technical Systems

Technology—Continuity and Change

Technology captures the lives and the imagination of most of us, whether we like it or not. Among those of us who grew up in the Space Age, it is a rare individual who did not rise early one morning to see a Mercury liftoff or huddle on a crowded schoolroom floor while the TV announcer described the exploits of John Glenn. Not so long afterwards, the new consciousness of the counterculture made it plain that technology was a mixed blessing. Ravaging the environment, controlled by a technocratic elite, undermining the unity of humankind and nature--the forces of technology had spiraled out of control.

Thus, unlike many academic subjects, technology is not one which requires scholars to convince us of its importance. The interpenetration of material artifacts throughout the structures of our everyday lives needs no evidence besides the telephone and automobile. Their importance in contemporary religious and political affairs is shown by the increasing use of the mass media in worship and electoral activities. Of the major social forces shaping our society and culture, none is so much discussed, admired, idealized, despised, and feared.

Perhaps the reason for such an intensity of reaction is based on a misunderstanding, namely the sense that technology is something new, with the potential to change and distress our social systems in complex and fundamental ways. Without disputing this potential for change, the notion of novelty is seriously mistaken. Here, scholarship has quite clearly shown the importance of technology in constraining the possibilities for societal development in all eras of sociocultural evolution (Lenski and Lenski, 1974). The earliest societies were just as surely shaped by the hunting technologies available to them as our contemporary societies are affected by their dependence on complex information and communication systems. The introduction of plant cultivation as a new mode of production changed these societies in basic ways, allowing, for the first time, the creation of a surplus, the establishment of permanent settlements, the accumulation of possessions, and the growth of inequality. It is arguable that such technology-induced changes were more far-reaching in their implications than those associated with the industrial revolution. Whether or not this is so, it is safe to say that technology has always impacted the social systems of humans in significant ways.

But there is truth behind the reaction to technology as well. If innovation has always been crucial to social development, it has never proceeded at such a pace. Indeed, the current rate of technological change is unparalleled and its social consequences are only beginning to unfold. Part of the explanation for this increase are simple continuations of conditions which have increased the rate of innovation since the horticultural period: an increase in the size of the population, increases in the stock of existing technological knowledge, and advances in communication and diffusion between cultures. Still, at least two features of industrial societies constitute new developments, institutionalizing innovation and guaranteeing its systematic production: (1) the linkage of technology with science and (2) the use of technology to achieve private and public organizational objectives.

In the early stages of the industrial revolution, new technology was produced by craftsmen and amateurs without scientific training. However, in the latter half of the nineteenth century, technology began to reflect the application of scientific principles, initiating the modern era of scientific technology (Rapp, 1981). The growth of technical education, the development of research laboratories in industry, and the establishment of the engineering disciplines encouraged this linkage (Noble, 1977). Daniel Bell has suggested that science-based technology is the principal source of both innovation and policy formation in "postindustrial" society (1973).

The other major factor increasing the rate of innovation is the use of technology to achieve public and private organizational objectives. With the development of capitalistic trade and the factory system of production, the search for new processes and products was stimulated by the incentive of private profit. Entrepreneurs who increased the efficiency of their production through the adoption of new techniques gained a market advantage over their competitors. Firms which diversified against dependency on a few products could enter new markets, guard against economic fluctuations, and invest profits in continuous growth. Technological innovation was recognized as one crucial element in this growth, first in the electrical and chemical industries, and later throughout the private sector.

Meanwhile, the military advantages brought about by investment in scientific technology were realized by all industrial nations. These acted as a stimulus to governmental support during peacetime as well. As a result of these interests, global support of science and technology stands at roughly $150 billion per year (Norman, 1979). Given the massive financial infusions, legions of research workers, and imponderably complex instrumentation which characterizes contemporary research and development (R&D) systems, it is no wonder this has been called the era of Big Science (Price, 1963).

Technology, then, has been a significant constraint and source of social change throughout the history of humankind, but its relationships with institutions of the modern world are new. Although this much is indisputable, only the broad outlines of these relationships have achieved a measure of consensus among scholars. We are still far from a complete understanding of contemporary technology, its sources, and its implications.

Approaches to Technological Innovation

One area within the general field of science, technology, and society has received a significant degree of attention over the years. The process of technological innovation--the development of new products and processes--is quite obviously a subject which intersects the interests of historians, sociologists, economists, policymakers, and industrial managers. A large number of studies have addressed this problem, generally taking one of two approaches.

The oldest and perhaps most obvious approach is a history of inventions, such as we find in the multivolume histories of technology or the "event-history" analyses examining the role of science in technological progress. This approach is useful for its highly detailed description of currents of intellectual influence and attention to the "micro" level of the innovation process. Much of the time, the focus is on individual inventors, their ideas, trials, and careers. Although such studies provide crucial evidence for general accounts of technological development, suggest hypotheses for systematic study, and qualify generalizations in important ways, they rarely go beyond the case study and the use of *individuals* as analytical units.

A second traditional approach focuses on the role of *organization* in the process of technological innovation. Ironically, one of the greatest individual inventors, Thomas Edison, played a significant part in the developments which led to this newer focus. For besides the electric light, the telegraph, and the inventions for which he is best known, Edison founded one of the first research laboratories in the United States.

The industrial research laboratory institutionalized key elements of technological change through a permanent staff of scientifically trained personnel, continuously employed in incremental inventive effort. In all classical treatments of technological dynamism the individual firm is the key unit in the productive process. Invention, innovation, and diffusion are three broad components of economic transformation brought about through the replacement of old technology (Schumpeter, 1950). These components are functions of firm-based processes. The efforts of specific organizations--guided by management decisions to invest in research and development, to innovate a product, to imitate rivals--are usually treated in terms of a stage model, e.g., idea generation, problem solving, and implementation (Utterback, 1971).

Relations between organizations enter into this process through the notions of competition and monopoly power.

A primary objective of this book is to propose a third approach, which combines elements of the first two, situates innovation in the contemporary context of "collectivized" science (Ziman, 1983), and focuses on the *system* of innovation as a whole. In part, but only in part, this involves a rejection of the previous views of innovation as incomplete and in some ways misleading approaches to the contemporary process of technological change. Both are misleading in treating innovation in an "atomistic" fashion. At best, the firm-based view is an "organization set" approach to the problem (Evan, 1966). Direct relations between a focal organization and other organizations are considered, but indirect linkages and potentially important relations *among* these organizations are generally ignored. The view is incomplete in treating firms as autonomous actors in conceiving and carrying an innovation to term, when much of the activity of knowledge production in contemporary societies is carried on in quite a different context. The event-history view, on the other hand, tends to focus on individuals and intellectual developments to the exclusion of social and organizational factors.

That technological projects vary in scope and character is recognized even by students of firm-based innovation. Differences between developments in size and complexity are likely to invalidate findings based on individual firms. James Utterback is careful to distinguish "very complex systems which involve many years to implement and resources far beyond those of a single firm" (1971: 77). Others indicate that large-scale innovations such as the Manhattan Project are poor models for industrial innovation (Schon, 1967: 40). Yet, in spite of this recognition there has been an almost complete lack of scholarly interest in large-scale technological systems. This is especially surprising in view of the acceptance by many economists of the importance of large firm size and governmental sponsorship of research owing to the increasing scope and complexity of modern technology (e.g., Galbraith, 1967).

In fact, large-scale technological endeavors pose critical issues both for students of knowledge production in advanced industrial societies and for policymakers seeking technological solutions to problems. As the scope of technological programs increases, the patterns and interrelationships characterizing project organizations become less like those which have previously been the focus of attention. New questions are raised for social scientific analysis and managers who seek ways of controlling these enterprises.

The remainder of this chapter provides a brief introduction to the subject of large-scale technological systems in two parts. First, three of the largest and most successful programs are described to give the reader a feel for the nature and scope of some relatively familiar cases of government-sponsored technological innovation. Next, several of the most important features of such programs are abstracted and presented in outline form. Here, their size, diversity, and formal organization are emphasized. The chapter concludes by summarizing the general argument of the remainder of the book using nuclear waste disposal and solar cell development as examples of currently important technological areas.

Large-scale Technological Enterprises

Throughout the history of civilization there have been attempts to employ technology on a large scale. The pyramids, the Roman system of roads and aqueducts, and the Suez Canal represent extensive state-funded enterprises. The Gothic cathedrals, the Crystal Palace, and the transatlantic telegraph cable were equally significant private undertakings. All these efforts were large-scale, interdisciplinary projects which required considerable expertise in design, engineering, and construction. Yet there were no extensive scientific underpinnings to these developments, and often the design work can be attributed to one individual, such as Ferdinand Marie de Lesseps (the Suez Canal), Cyrus Field (the telegraph cable), or Joseph Paxton (the Crystal Palace). That is, such large-scale developments did not involve coordinated *research* to any significant degree. Not until the twentieth century are there large-scale enterprises which produce and utilize scientific technology through the coordination of multiple organizational components transcending state and private boundaries. The prototypical instance of such an enterprise was the Manhattan Project, which developed the first atomic weapons during World War II.

The Manhattan Project

The Manhattan Engineer District was formed in August 1942, on the recommendation of President Franklin Roosevelt's S-1 committee. The committee had estimated earlier that year that a crash program costing $100 million might produce an atomic weapon by 1944 (Lamont, 1965). At its completion, the project had cost $2 billion, involved 600,000 individuals, and leased more land than the state of Rhode Island (Groves, 1962; Robinson, 1950). Beginning with a meagre knowledge of nuclear fission and Enrico Fermi's experimental demonstration of a chain reaction, a vast organization of theoretical and experimental physicists, chemists, mathematicians, biologists, chemical engineers, ordnance experts, metallurgists, and other technical personnel was assembled.[1] This technical component was joined with an administrative component of both scientists and military personnel and with mechanics, electricians, construction workers, and others necessary to implement the project.[2]

Material and manpower resources were drawn from both public and private organizations. The Army Corps of Engineers served as an essential source of engineering expertise, but other governmental agencies were also involved at various stages of the project: the U.S. Employment Service, the Office of Education, the Bureau of Mines, the Bureau of Standards, and the Ballistics Research Laboratory, among others. Three primary organizations were created explicitly for the purposes of the project--Oak Ridge (which operated three uranium production plants), Hanford (the site for plutonium production reactors), and Los Alamos (bomb design and assembly). All were operated by contractors and would eventually become part of a system of national laboratories owned by the Atomic Energy Commission (AEC). Numerous university scientists were recruited to these sites, although a few continued research in the academic environment. Three universities--Chicago, Columbia, and California--received certificates of merit after the war for their participation. Besides du Pont, Union Carbide, and other prime contractors, a host of private firms contributed labor, expertise, and technology to the project.[3] The gaseous diffusion plant at Oak Ridge, for example, was the combined effort of basic research at Columbia University, a design by Kellex (a subsidiary of Kellogg Company), a builder (J. A. Jones Company), and an operator (Union Carbide). Other firms such as Allis Chalmers, Bart Laboratories, and the Chrysler Corporation added essential equipment and production processes (Groves, 1962). The entire project was characterized by formal organization, task differentiation, and clear lines of authority, the outline of which is shown in Exhibit 1.1.

Exhibit 1.1 Organization of the Manhattan Project*

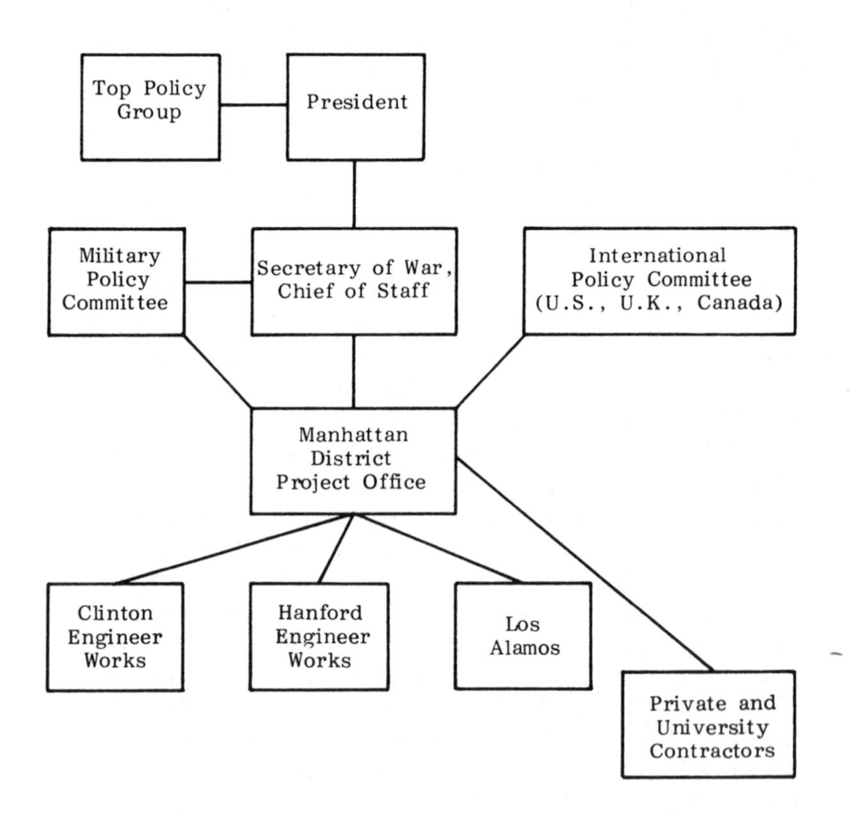

*Adapted from Leslie Groves, *Now It Can Be Told*. New York: Harper and Brothers, 1962.

With but a single overarching objective--the creation of an atomic bomb--the Manhattan Project must be judged effective. The outcome of the program brought the war against Japan to an abrupt conclusion. In terms of other criteria, however, the modern observer may not be so quick to pronounce it a complete success. Vast cost overruns, conflicts between scientists and nonscientists, deficiencies in planning and management, sabotage, high turnover and absenteeism, and the production of dangerous waste products were readily overlooked given the spectacular achievement of the primary aim. Yet it is difficult to overstate the importance of the Manhattan Project as a model for subsequent large-scale technological enterprises.

Polaris

Following World War II, government support for science and technology increased rapidly. Science had proven its effectiveness in military applications and all branches of the armed forces hastened to utilize high technology in the development of advanced weapons systems. The Polaris system developed by the Department of Defense (DOD) displays the same interdisciplinary, hierarchical, and multiorganizational features we have seen in the Manhattan Project.

The Fleet Ballistic Missile Program, better known as Polaris, was an effort which began in the late 1950s to develop a submarine-launched intercontinental ballistic missile. Managed by the Navy Special Projects Office, the Polaris missile and forty-one nuclear submarines were developed at a cost of over $10 billion three years ahead of schedule. A competitive, decentralized system of private aerospace and defense contractors supported by a large cadre of university researchers and consultants was marshalled by the special projects office in conjunction with naval technical bureaus and operating forces. Aside from the technical innovations, a number of managerial systems were introduced, including the Program Evaluation and Review Technique (PERT), heralded as the key to successful coordination of such large enterprises, but of unproven effectiveness (Sapolsky, 1972).

Apollo

During this same period, the federal government diverted massive funds to space research in response to the Soviet launching of Sputnik. Like the Polaris missile and the Manhattan Project, the Apollo mission was widely acclaimed as a technological success story. Its principal aim, landing a man on the moon, was achieved within the planned time frame and, discounting leftover hardware, within the original cost estimate of $20 billion. Based on the financial and manpower resources involved, Apollo has a legitimate claim to be the largest technological enterprise in history (Seamans and Ordway, 1977). At its peak, the National Aeronautics and Space Administration (NASA) employed 411,000 persons (only 8.3% were government employees), 300,000 of whom were directly involved with Apollo. In terms of organizations, over 200 universities, 80 foreign countries, and 20,000 private contractors (2,000 prime contractors and 18,000 lower-level contractors) helped to launch the moon flights. As in the Polaris project, Apollo managers utilized a variety of state-of-the-art management techniques, including phased project planning, systems management, configuration management, and PERT.[4] Further, both Polaris and Apollo enjoyed widespread and unremitting public support, including relatively unlimited funds from their respective departments. Such resource abundance also characterized the earlier Manhattan Project, though within a context of absolute secrecy.

Characteristics of Technical Systems

Unlimited resources and public support are the exception and not the rule where technical systems have been concerned, yet these historical cases all exemplify the essential characteristics of most large-scale technological enterprises. It is convenient to introduce a concept which will apply to the entire range of such entities. We may then describe their characteristics in a more systematic fashion. A *technical system* may be defined as a centrally administered network of actors (organizational as well as individual) oriented toward the achievement of a set of related technological objectives.[5] It is, in essence, an organization for producing innovation, an entity which specializes in collective problem solving. By "technical," the use and production of the kind of specialized knowledge characteristic of science and technology is implied; by "system," both functional and structural interdependence of components is suggested.

Exhibit 1.2 summarizes several of the characteristics which distinguish technical systems most clearly from smaller analytical units such as industrial firms or basic scientific specialties. These features are discussed under the general headings of size, diversity, and formal organization.

Size

The features of size and complexity are significant for the contrast between technological innovation in the firm-based model and innovation in technical systems, principally due to the increased coordination and control require-

Exhibit 1.2 Characteristics of Technical Systems

A. LARGE SIZE

1. Funds
2. Personnel
3. Organizations

B. COGNITIVE COMPLEXITY

1. Basic, applied, and developmental research
2. Multidisciplinary

C. OCCUPATIONAL PLURALISM

1. Researchers
2. Program managers
3. Policymakers
4. Public-interest advocates

D. SECTORAL DIVERSITY

1. Governmental agencies
2. National laboratories
3. Private firms
4. Universities
5. Public-interest groups

E. FORMAL ORGANIZATION

1. Central administration
2. Hierarchy of authority
3. Division of labor

ments involved. Although most technical systems would hardly match Apollo in terms of size, all are multiorganizational, often comprising hundreds of organizations in various roles. Many of these organizations are quite peripheral to the central objectives of the system, such as subcontractors, universities with lone researchers, and federal agencies with minimal interests, yet the core of the system will still consist of several dozen organizations. Of course, funds to support such an operation might come from the federal government, private enterprise, foundations, or consortiums, but they will generally be more than even the largest private firms can finance in the absence of joint efforts.

Diversity

The diversity of technical systems is expressed in terms of occupations, disciplines, and sectors. The demanding, often staggering, technical objectives make every technical system an interdisciplinary enterprise, calling upon the cognitive resources of many different scientific and engineering specialties. "Core" fields (e.g., physics in the Manhattan Project) are often supplemented by an array of related, but less central fields (medical scientists and biologists were called in to study the effects of radiation on humans).

Occupational pluralism is the inevitable result of task differentiation within a technical system. In addition to scientists and researchers a large administrative component is needed to perform integrative and control functions. Workers of all varieties from electricians to machine operators are required to carry out construction and/or production of the technology. Although this report focuses on the earlier stages of innovation in technical systems, the organization and management of implementation are certainly as critical to the overall success of the system.

Sectoral diversity is another aspect which will concern us in an analysis of technical systems. Indeed, in many respects it is the most significant dimension of all. A "sector" is an organizational type with a distinctive set of structures, priorities, and outputs (Marcson, 1972). A basic type of sectoral differentiation observes the fundamental economic and political dichotomy of public and private. This is an important dimension, but is too crude to represent the types of organizations found in technical systems. National laboratories, academic institutions, and private firms (and often nonprofit research institutes), all distinctive structural entities, serve the research function while governmental agencies (operational, regulatory, oversight, and advisory), public-interest groups, and professional associations influence policy and administrative functions. The precise number of sectors may vary from one system to another, but will usually include national laboratories, universities, private entities, and governmental agencies at minimum. More importantly, the *distribution* of involved organizations among these sectors will vary significantly from system to system, with important implications for the nature of the system and the innovation processes it supports. This distribution and its consequences are a principal theme in the chapters that follow.

Formal Organization

The size and complexity of technical systems are the main reasons that some type of formal organization is necessary. This feature also serves to distinguish technical systems from firm-based innovative modes and certainly from the informal and collegial organization of basic science. Requirements for control and coordination are generated--a multitude of "interfaces," or component compatibility questions, scheduling of distinct organizational developments, trade-offs between competing technical alternatives. Of course, it would be ludicrous to suggest sending a man to the moon without some plan for accomplishing this objective or an agency which could be held accountable. This plan, and the structures established to secure its implementation, is the formal organization of the technical system, a *sine qua non* of its existence.

The federal government provides the formal planning and administrative structures for most technical systems. We have suggested that the massive resource demands for many developmental projects are one reason for this. An equally important factor is the role of the state in advanced industrial societies (O'Connor, 1973).

Economists have long recognized the imperfections of the market in eliciting optimal levels of spending on R&D. Where firms cannot appropriate the returns of their investments, they are unwilling to generate internal R&D (Mansfield, 1971). A second feature of capitalist economies is their failure to provide collective goods (e.g., clean air and water). In advanced industrial societies it has fallen to the state to introduce corrective mechanisms, embodied in regulatory, fiscal, and monetary policy. One such mechanism is the establishment of governmental systems in the search for technological solutions to social problems.

The simplest form of system would be the appropriation by government of a pool of grant monies and their distribution through a central office. In more developed form, one or more administrative offices would be established, usually within an existing agency, a detailed development plan formulated (either by the office or its contractors), and a complex series of tasks allocated to university and private contractors, with agency laboratories playing more or less central roles.[6] Thus a relatively centralized administrative structure, a hierarchy of authority, and a division of labor--all features of classical bureaucracies--characterize technical systems.

Formal organization (in both its planned and bureaucratic aspects) serves as a common ground linking innovation in firms and technical systems. A pool of capital, technology, and labor resources is set at the disposal of a centralized management group, which serves as final arbitrator in debates over technical alternatives, allocates manpower and funds to tasks, and assembles interdependent components to produce material artifacts and procedural systems.

But the *interorganizational* features of their formal structure argue the special nature of technical systems. This is because the single firm has exclusive control of its divisions and component units and centralized direction is relatively effective--or at least unproblematic. However, a technical system consists of many organizations in different sectors, each operating in relatively autonomous fashion. The control and coordination of such entities is inherently difficult, leading to a different kind of conceptualization--that of a "loosely-coupled system" (Weick, 1976). An interorganizational research network organized by a central administrative component is not, then, simply a large-scale organization (although the analogy is helpful in some contexts), but a grouping of semi-independent units whose participation and cooperation are in principle problematic. Indeed, the conflict between organizational and systemic objectives is a central problem for both the analyst and the manager of such systems.

Objective of the Study

Though the significance of technical systems is not in dispute, there is a lack of systematic study of their structures, processes, and relations to the environment. In general there are histories, case studies, and descriptive accounts such as those of Manhattan, Apollo, and Polaris, but few quantitative, comparative analyses. Notably, there is already at least one managerial text on the subject (Sayles and Chandler, 1971). Allan Mazur and Elma Boyko studied large-scale ocean research projects using interviews with key participants. Although these were relatively smaller in scope than technical systems, a division of labor, in which separate administrative roles were established, seemed to be associated with success, as well as autonomy from the funding agency (1981). Kenneth Studer and Daryl Chubin (1980) demonstrate the importance of national laboratories in the cancer research system. The literature on the development of scientific specialties, while providing suggestive models of social and cognitive development, deals exclusively with informal or unorganized collectivities, representing a stark contrast with the formal organization of technical systems (Mulkay, 1980). The several speculative, programmatic statements, including L. Vaughn Blankenship (1974), A. Szalai (1979), Bruce Hannay and Robert McGinn (1980), and Mel Horwitch (1979) have utilized examples, but no systematic data.

The purpose of this report is to introduce a systematic approach to the study of technical systems. In doing so, an argument is offered relating the use of the technology under development to the process of innovation. One system, nuclear waste management, has developed in response to a problem generated by technology itself, the operation of nuclear reactors. The other, solar photovoltaic cells, is an attempt to promote rapid commercialization of an alternative energy source. The broad objectives of the two systems condition the observed patterns of funding and organizational involvement--what we may call their "interorganizational structure." This in turn affects the nature of the innovation process. An attempt is made to determine the roles diverse sectors play within a system, the degree to which individual actors in each sector have social ties with one another, and the influence of such ties on the innovativeness of research workers.[7]

It is important to state at the outset what this study does not do. The model of technical systems presented in Exhibits 1.2 and 3.2 suggests a number of significant problems for investigation, many of which are passed over quickly. In particular, the formation of fields through the interaction of organizations in various national policy domains receives only a descriptive treatment. Likewise, the role of governmental regulatory agencies and public-interest groups, the relationships among numerous occupations and professions in the innovation process, and the implementation of solutions are not given the space which these topics warrant. Instead, the focus is on the research component of technical systems, communication patterns, and their consequences.

The report is organized in two parts, corresponding to the qualitative and quantitative components of the argument. Part One summarizes the results of historical and organizational analyses, drawing largely on published and unpublished histories, interviews with numerous informants, and a great many program summaries and progress reports published by governmental agencies. Part Two presents the results of a structured interview with 297 participants in radioactive waste and photovoltaics, focusing on communication and performance. It is possible to read Part One without continuing if one is willing to be satisfied with a macro-level account of the differences between the two systems. It is much less desirable to read Part Two in isolation, because the interpretation of the survey results depends on the nature of the systems described in Part One.

A preliminary word about the nature of the study may help to place it in context. Since the study is frankly exploratory rather than confirmatory, hypotheses are not "tested" in any rigorous sense. It is obvious that cross-sectional data are a poor excuse for the kind of information necessary to establish causality. Further, since only two cases are examined, the generality of the results presented is very much open to question. Technical systems exist in health, agriculture, defense, space, and other fields besides energy. Their size and form of organization may make them different in fundamental ways from nuclear waste and photovoltaics.

An approach possible within these constraints is to build an argument which accounts for interesting features of the technical systems under examination here and which may be applicable to other systems. That is, the report presents an account rendered plausible by limited evidence. In brief, the account goes as follows.

The Argument in Brief

In Western industrial societies the state seeks to provide collective goods and stimulate the provision of private goods for its citizens. Specific technologies are utilized in the service of either collective or private goods. This use shapes the distribution of resources and organizational actors which develop the technology and, hence, the process of technological innovation. In systems oriented toward the provision of collective goods, the government has a relative monopoly on developmental resources (funds and, to a lesser degree, personnel) and a much greater degree of control over the process of innovation. In such systems one observes a fundamental shift in the nature of the communication relevant to scientific performance, with program managers playing the crucial role rather than members of the scientific community. In essence, we argue that Big Science has its origins in dependence on government, not industry.

The evidence is marshalled in a series of four steps. In Chapter Two, nuclear waste disposal and solar photovoltaic cells are introduced as examples of the use of technology for collective and private values, respectively. The historical development of nuclear waste and photovoltaics demonstrates the importance of political and economic factors in the emergence and growth of technical systems. That there is nothing inherent in the technology which makes it "public" or "private" is clear from the case of photovoltaics, which was initially developed in the service of space flight. That political controversy can generate the conditions for the growth of a large-scale technological enterprise and be summoned in the interests of diverse disciplines is nicely illustrated by nuclear waste. A sense of the complexity of the scientific and technological tasks involved is provided by a description of current research problems in these fields. Lastly, evidence is presented which suggests that the allocation of research effort may be planned to a greater degree in the nuclear waste system.

In Chapter Three the interorganizational structure of the nuclear waste and photovoltaic systems is described. In many respects, such as size, organizational composition, and administration, the systems are similar. Hierarchical authority structures and complex management systems are employed to coordinate the efforts of organizations in private, academic, and federal sectors. The generating milieu and the greater complexity of the technology are associated with a larger administrative component in radioactive waste and a greater number of actors in the environment of the system. Most importantly, though, the distribution of funds and sectoral involvement is shown to favor the private sector in photovoltaics, the public sector in nuclear waste. The monopolization of research funds by the federal government and the noncompetitive nature of laboratory research are evidence of a significantly different interorganizational structure in the radioactive waste system.

The consequences of this structure are explored in the quantitative analyses of Chapters Four and Five. Chapter Four uses information on the social relationships among system participants to show that while the government provides the integration function for both systems, they are quite different in terms of the pattern of relationships characterizing the research sector. In nuclear waste, there are strong linkages between and within government and national laboratory sectors--the "public core." Academic and private sectors are less important to communication within the system. In photovoltaics, a more uniform pattern of ties among the research sectors is evident, although this is not necessarily true of professional exchanges.

Finally, Chapter Five asks the question: what kinds of linkages are associated with high levels of research performance in each system? Based on the preceding argument, a hypothesis is formulated suggesting that the nature of nuclear waste technology as a collective good and the relatively complete dominance of the system by the federal government will result in the greater importance of ties with the public sector. In photovoltaics, the greater involvement of the private sector and the investment of private research monies reduce dependence on government and should lead to the greater importance of contacts with other researchers, a pattern typical of basic, academic science. The results of a multivariate analysis support this notion.

The implications of the findings are significant for both the sociological study of science and the management of innovation in technical systems. In systems which involve a "collective good" the monopolization of research resources gives administrators and managers a degree of control over the research process not possible in systems which are "competitive" with respect to private-sector involvement. In this sense, it is dependence on government, not industry, which is associated with a change in the nature of science.

Part One

Historical and Organizational Context

Chapter 2

Collective Goods and Private Profits

Energy and Politics

During the early 1970s, largely as a result of the oil embargo, policymakers in the United States and elsewhere adopted a definition of energy as a problem which required attention at the national level. Government spending on energy R&D increased from $500 million to $3.5 billion from 1972 to 1979 (Norman, 1979). Many forms of energy were examined, but two of the most closely scrutinized were nuclear and solar technologies. Ideally the range of options would have been assessed in rational cost-benefit terms, and doubtless they were, in contracted evaluations and scholarly analyses. But much of the energy debate took place in another arena, in which environmental politics, fiscal policy, and a myriad of structural and rhetorical elements played a role.

The federal government, which had invested heavily in the nuclear industry since its inception, was led to consider, at long last, that the wastes produced by the fission process might constitute an insurmountable problem as the industry expanded to meet national energy needs. Despite a number of technical advances in the treatment and isolation of radioactive waste products, no solution was defined as acceptable by the majority of actors in the process--and certainly none was so defined by residents near the anticipated disposal sites.

A veil of controversy has shrouded the nuclear technologies both in their military and commercial aspects, and the technology of disposal is no exception. Indeed, the antinuclear movement has adopted the waste issue as a major tool in the debate over nuclear power.

The solar industry, though it takes a much different shape than the nuclear, began to receive considerable emphasis during this same period. Clean, safe, reliable, and utilizing a "free" energy source, photovoltaic cells were characterized as the antithesis of the fission reactor. The solar technologies were idealized, legitimating the technologists who sought an entrepreneurial answer to the energy crisis (reduced dependency through inventiveness) and the environmentalists who sought benign alternatives to pollution and risk. The optimistic forecast an America of rooftop arrays and free electricity.

It would be simplistic to characterize technical systems based on public perceptions of the technology. The emerging milieu seems at first to suggest a distinction between nuclear waste and solar cells based on a conception of the former as a social problem and the latter as a social benefit. Indeed, the fact is crucial for an analysis of the relations between technology and social movements, conditioned as they are by definitions of the public interest. Its relevance is not so clear for the process of innovation within technical systems, the central concern of this report.

Solar photovoltaics and nuclear waste are two active fields of energy research and development. Both began in the 1950s based on previous scientific and technological developments. Both are extensively funded by the federal government, with massive yearly appropriations and hundreds of research workers. Both rely on similar forms of formal organization to allocate funds to research tasks and coordinate numerous aspects of the endeavor. Both utilize the resources and expertise of diverse sectors--private, academic, and public. Thus, in spite of their polar affiliations with the environmental movement and the energy policy process, they have a common objective--the production of technological innovation. In this they are quite similar.

Types of Technical Systems

To what degree may systems concerned with innovation be seen as the "same?" There exists no established classificatory scheme for technical systems. Clearly, a comparison based on the nuclear/solar dichotomy would be of limited interest outside of the energy field. A typology based on the nature of the output of the system, the economist's contrast between private and collective goods, is much more useful as a general analytical tool.

Private goods are commodities or values that can be possessed by an individual. Because they come in *discrete* units they are available to some and can be withheld from other members of a society. A collective good, on the other hand, possesses a "common value," such that "if any person X_i in a group $X_1, \ldots, X_i, \ldots, X_n$ consumes it, it cannot feasibly be withheld from the others in that group" (Olson, 1965: 14). The essence of a collective good is its *indivisibility*, a property which makes it infeasible to prevent one member of a society from obtaining the benefit of the good once the society as a whole has acquired it.

This conceptual tool allows us to distinguish between technical systems whose objectives are to provide collective goods and those whose objectives are to provide private goods. Weapons systems, environmental quality, and space exploration are examples of collective goods because their benefits cannot feasibly be withheld. Nuclear waste disposal is also a collective good because the prevention of radioactive contamination of the environment is a benefit to all members of society. Telecommunications devices, automobiles, health technologies, and agricultural techniques are private goods due to their divisibility--they must be paid for to be used. Photovoltaic cells are available only to those who can afford them and hence are private goods.

There is nothing about the technology itself which determines its collective nature. Nuclear reactor technology is a private value, inasmuch as electricity can be withheld from anyone who does not or cannot pay for it. But waste disposal is a collective value owing to the public nature of its benefits. Indeed, we will soon see that photovoltaic technology itself was developed for the collective benefits it could provide.

The collective/private distinction is important because of its implications for the involvement of various sectors in technological development. Clearly, it is not the case that government simply intervenes in areas where a technological fix will provide a collective value. Nuclear reactor technology is itself a prime example of another kind of involvement, wherein the state supported research and development, provided large capital outlays, and absorbed uncertainties in the development of the nuclear industry. In the case of a private good, the state "stimulates"; in the case of a collective good, the state "provides." Justification of state intervention may utilize similar--though not identical--ideological resources in both types of system. In both cases the public interest is purported to be the central issue, with the state carrying out its functions of ensuring economic productivity, national integrity, and quality of life.

When will the government step in to develop a technology? Economic theory suggests that the market mechanism is often deficient in allocating resources to R&D due to the inappropriability of returns on investment and the risks involved in developing nascent technologies (Schumpeter, 1950; Kamien and Schwartz, 1975; Ganz, 1980). The level of investment which might be optimal from the standpoint of the wider society (say, to develop a nonpolluting energy alternative) may be much greater than private firms devote to the technology for a range of reasons. Foremost among them is that the knowledge upon which new technology depends is often just as much an "externality" as the wastes which may be its byproducts. Firms may find that a reliance on the R&D of others is more profitable than investing their own resources. Too, American management is often characterized by its short-term profit horizons in which investment for new technology is viewed as a low priority. Further, firm size may simply be insufficient to sustain the massive levels of funding necessary to develop some new products. In any of these cases state decision makers may come to adopt a strategy justifying intervention through the establishment of procurement programs or incentive plans. The creation of a technical system--organizing a research and development program to provide a technological solution--is another form of response.

The traditional concept of externality refers to byproducts of industrial activity, such as waste or pollution, the costs of which are not borne by the producing agents. Their removal is a collective good, the provision of an uncontaminated environment, and is generally defined as the responsibility of the state. Other activities, such as the defense of the country, the exploration of space, and the accumulation of basic scientific knowledge are equally held to be appropriate functions of the federal government. The absence of a commercial market seems to be a common denominator in these areas, but such an absence is at least in part based on policy processes and regulatory activity.

What is at stake here are the vastly greater resources of government and the influence of interest groups, lobbies, and other actors in the policy and budgeting process. The origins of technical systems have not received much systematic attention. However, Wolfgang van den Daele, Wolfgang Krohn, and Peter Weingart have proposed a three stage model for the emergence of "hybrid communities," quite similar to technical systems, beginning with the transformation of a social problem into a political program (1977). This political program is then linked to a set of technological objectives, transforming it into a science policy program. Finally, concrete research planning begins, often with "transfer organizations" which undertake the coordination and funding of the research effort. This model is intended to apply to all forms of state-organized technological efforts. However, by distinguishing collective and private goods, the profit potential of the output emerges as a crucial issue. The principle of *cui bono*[1] suggests that a collective good such as nuclear waste management is likely to have only one customer--the state--while a private good such as photovoltaic cells, may have significant commercial potential. Private goods, then, are likely to generate investments from the private sector, such that the burden of R&D will not fall exclusively on government.

To this point we have distinguished between technical systems based on whether the output of the system is used as a collective or private good. Nuclear waste disposal and solar photovoltaic development have been introduced as cases typifying the distinction, with the former representing a collective good. We turn now to the historical development of each field, showing their emergence as distinctive technological enterprises connected to a variety of wider social and economic concerns. As *technological* fields, it is important to have some notion of the problems which they are organized to solve, so part of the discussion deals with these issues in a general, nontechnical fashion. The final part of the chapter presents evidence on the distribution of research work in each field which suggests the greater centralized control over the organization of research in nuclear waste. It is the first piece of evidence that systems concerned with collective goods may function differently than those concerned with private goods.

Radioactive Waste Disposal

January 7, 1983, marked the passage of the Nuclear Waste Policy Act of 1982, a key date in the history of nuclear waste disposal. For the first time, Congress had established a process and schedule for the development of nuclear waste repositories, a step towards the resolution of a problem which had been growing in magnitude since the first radioactive wastes were produced during the last years of World War II. But the research and development which provided the basis for the decision have been in progress for over twenty years--and it will not stop with the construction of the first repository, to begin in 1987.[2]

Care and Feeding of Nuclear Waste

Radioactive waste, simply defined, is waste material contaminated by radioactive isotopes. Most is produced through the operation of nuclear power reactors and the manufacture of nuclear weapons. Other sources of waste are research investigations, medical diagnostics and therapy, and the mining of uranium ore. Historically, the nature and type of waste have been crucial to its management, particularly in regards to the distinction between military and commercial waste. High-level wastes (HLW), transuranic wastes (TRU), low-level wastes (LLW), uranium mine and mill tailings, and gaseous effluents all involve separate treatment and management decisions. In addition, spent fuel from commercial power reactors has sometimes been considered a waste form, sometimes simply a pretreatment phase, depending on the status of the reprocessing issue. Quantities of existing waste were inventoried in 1979 by President Carter's Interagency Review Group on Nuclear Waste Management (IRG): 9,480 thousand cubic feet of high-level waste, 1,223 kilograms of transuranic wastes, 2,300 metric tons of spent fuel, 66.6 million cubic feet of low-level waste, and 140 million tons of uranium mill tailings. The group also estimated the magnitude of the management task involved by projecting "lifetime" requirements for two scenarios (low and high nuclear capacity) to the year 2000: 4-8 geologic repositories, 3-14 spent-fuel storage facilities, 440-1,650 facilities decommissioned, and 1,400-3,200 trips with high-level waste (Interagency Review Group, 1979: 12).

The principal concern with respect to radioactive waste is the biological hazards of ionizing radiation, including carcinogenic, mutagenic, and teratogenic effects as well as death through exposure to or uptake of radionuclides (Lipschutz, 1980). The most critical scientific fact about radioactive material is noncontroversial: "there is no method of altering the period of time in which a particular species remains radioactive, and thereby potentially toxic and hazardous without changing that species" (IRG, 1979: 9). Fission products, formed by splitting of heavy elements, generally decay to innocuous elements within the first thousand years. However, transuranic elements such as plutonium decay slowly, leading one observer to claim that it is almost impossible to make a rational case for any specific containment period from one hundred thousand to ten million years (Gera, 1975: 14). In essence, the technical system for radioactive waste disposal is oriented toward the search for a method of isolating these materials from the biosphere for extended time periods, removal from the earth, or transformation to nonradioactive forms.

A History of Management

The history of radioactive waste management may be divided into two periods, marked by the breakup of the Atomic Energy Commission in 1974. There is general agreement among those who have reviewed the history of waste management that (1) the problem received relatively little emphasis until the late 1960s; (2) the Atomic Energy Commission was in large part responsible for the loss of credibility by the federal government due to shortsighted management decisions; and (3) the technical system which exists today encompasses many of the same actors and policy issues it did in the 1950s, but its character has been changed by the introduction of new social and cultural elements.

Of course, it was recognized in a general way from the very beginning that waste would be produced by the defense establishment. Nonetheless, the AEC did not address the problem publicly until 1949, and then in terms of "handling" rather than "disposal" (Hewlett, 1979). The problem, wrote Carroll Wilson, the first general manager of the AEC, was the relatively low prestige attached to nuclear waste research: "Chemists and chemical engineers were not interested in dealing with waste. It was not glamorous; there were no careers; it was messy; nobody got brownie points for caring about nuclear waste. The AEC neglected the problem" (1979). A current DOE informant expressed the same sentiment:

> Nuclear waste management starts with a social stigma. [In the] 1950s and 1960s the good people worked on reactors and enrichment, not waste.

It has been estimated that in its early years the AEC spent about $5 million per year on wastes (Shapiro, 1981: 14), about 1% of current annual expenditures.

A milestone in the debate over technological alternatives was the 1957 report of the National Academy of Sciences (NAS) which recommended the disposal of high-level liquid waste in bedded salt formations. Designated the "reference disposal concept," it involved a commitment to spent-fuel reprocessing, solidification of wastes, and packaging before emplacement in mined rock salt cavities (Greenwood, 1979). Although other disposal concepts have been studied, it has remained the preferred option well into the late 1970s.

During the period from 1957 to 1965, significant technical innovations were made in the field of waste treatment--in particular, volume reduction through a process of calcination (Willrich and Lester, 1977). But the AEC, rather than following up the NAS proposal, sought to dispose of wastes at the point of generation. Leaks began to occur as early as 1957, though claims were made that the tanks could last for decades. Although the calcination process was available, a decision was made to allow the liquid waste to evaporate into a damp salt cake, one of the costliest and least justifiable decisions of the period from a technical standpoint (Willrich and Lester, 1977; Ford Foundation, 1977).

Public criticism of the AEC began in the early 1960s, but interim storage continued to take precedence over ultimate disposal. A highly critical NAS report on storage practices completed in 1966 was suppressed, prior to the disbandment of the committee itself (Boffey, 1975). Continued leaks, a major fire at the Rocky Flats Plant in 1969, and the sudden collapse of plans to build a prototype repository at Lyons, Kansas, in 1972 led to increasing pressure for the development of an integrated waste policy (Carter, 1977; Lipschutz, 1980).

In essence, this policy involved reprocessing and solidification prior to permanent disposal in a bedded salt repository. Defense wastes were to be stored on site. The problem of radioactive waste *management* was and continues to be the interdependence of elements, such that decisions or failures with respect to one area have implications for others. Difficulties during this period ranged from the cancellation of the Lyons repository, to the leakage of defense wastes, to the termination of three reprocessing plants in either the planning or operational stages (Hewlett, 1979).

However, a well-publicized leak of 115,000 gallons of liquid high-level waste at the Hanford reservation in 1973 was the most salient event in bringing the nuclear waste problem to the attention of the public. The leak was particularly disturbing since it went undetected for fifty-one days even though the tank level was monitored daily. In public opinion surveys from 1960 to 1973, no respondents brought up the subject of nuclear wastes when asked for reasons why they were opposed to nuclear power. Since 1973, when presented with a list of disadvantages, radioactive waste has been the issue most frequently picked (Nealey and Hebert, 1983).

It should be emphasized that a lack of technological innovation is not generally considered to be the principal problem during these years. The lack of a solution to the waste issue must be attributed to the failure of the technical system as a whole. Informants in the earth sciences, speaking of the period, laid the blame to the organizational climate of the AEC:

> The AEC was a juggernaut. The people were too involved and the momentum such that alternatives and problems were swept away without consideration.
>
> Most AEC people seemed very condescending. [They had a] "leave it to us" attitude.

One former manager with the commission expressed a quite different opinion of the period:

> I'm cynical about government activities [now]. [There are] too many reorganizations. The AEC had a small staff, well organized. I worked in production and we worked closely with R&D.

It seems clear that the relatively smaller size and the secrecy surrounding many nuclear activities contributed to the formation of a close-knit solidarity group which antagonized many state and local actors whose cooperation was needed. It might be said that the focus on technical innovation proved to be a major impediment. Although the breadth and depth of technical studies has been an issue recently, the perception that radioactive waste disposal is not primarily a technical problem is virtually universal. The view of one informant that "nuclear waste is political--the science is peripheral" was fairly typical.

For many years social and political aspects of siting, public participation, and involvement of independent experts were neglected. When pressures from outside the technical system came to constrain and challenge its operation, social and institutional issues were then addressed. But not before the character of the system, the pace, and the direction of technological development had changed: "The ability of the responsible officials to understand sufficiently what was happening and to take appropriate action to rebuild confidence was inadequate. The result was a growing paralysis of and public hostility and distrust toward the government's program, each of which fed one and reinforced the other" (Greenwood, 1979: 6-7). Richard Hewlett, chief historian of the Department of Energy (DOE), suggests that lip service was given to nontechnical factors such as public acceptance but there is "no evidence at all that attention was given to such matters as social, cultural, or psychological phenomena that might serve as constraints in implementing technical solutions" (1979: 4). Even the formal organizational arrangement was problematic. The separate organizational structures for the management of defense and commercial wastes were perceived as evidence of an uncoordinated and inadequate approach, further reducing the credibility of the AEC.

The years surrounding the split-up of this agency were also a period of increasing scientific *dissensus* on waste management practices. Alluding to technical questions, Irvine C. Bupp observes "it was easier in 1978 than in 1968 to find seemingly qualified technicians and scientists prepared to give contradictory answers" (1979: 122). The public concern was buttressed by technical arguments regarding the inadequacies of past practices and present knowledge. Initially, technical problems of waste disposal were not considered large, and the technical system devoted to waste management was correspondingly small. The dominance of economic over safety considerations, adoption of temporary expedients, managerial misjudgments, and insufficient resources for both technical and nontechnical studies contributed to the sequence of events undermining public confidence in federal waste management and forced the growth of a larger and more complex technical system.

New Directions

Up to 1974, the budget for waste management had never exceeded $61 million. In 1975, as the Energy Research and Development Administration (ERDA) took charge of radioactive waste research and management, the budget increased to $94 million. By 1977, when the current Department of Energy was created, this had increased to $230 million (IRG, 1979). As the nuclear waste issue came to play an ever larger role in discussions of the viability of nuclear power, the technical system received an influx of resources, accompanied by organizational shifts.

The breakup of the AEC and creation of ERDA allowed a policy review. Without its old regulatory functions, its decisions acquired greater credibility. Nonetheless, most of the personnel involved were drawn from the old agency and ERDA followed similar approaches to storage and disposal of wastes (Hewlett, 1979: 31-32), including the Retrievable Surface Storage Facility (RSSF). This idea has frequently been proposed and criticized (and in this case, rejected) on the grounds that it is only a temporary solution. Management of commercial and defense wastes was finally centralized by ERDA.

Three other innovations may be attributed to ERDA. The first was the notion of "multiple barriers" to be emplaced between the waste and its environment, including a high-integrity container in which to seal the solidified waste. Currently, the multiple barrier notion involves the waste form, stabilizer, container, overpack, migration retardant, and backfill (Carter, 1979). Secondly, a decision was made to make a nationwide search for suitable repository sites in a variety of geographic locations and geologic formations. The third innovation was ideological. Waste management debates had always involved discussions of the relative benefits of retrievable versus permanent storage, a clear tradeoff between the costs of surveillance and the possibility of technical problems developing at a permanent facility. Now managers began to speak of "terminal" storage. In Richard Hewlett's words: "The word terminal solved the old dilemma of irretrievability. Because all the wastes were to be sealed in containers, they could be placed in geologic storage in either a retrievable or irretrievable mode. By backfilling and sealing, the retrievable mode could be changed to essentially irretrievable. Thus, geologic sites could be called terminal storage facilities rather than retrievable or irretrievable" (1979: 34-35). Thus, the establishment of the National Waste Terminal Storage (NWTS) program in 1976 was more than a proposal to evaluate sites in thirty-six states. The naming of the program itself was a response to pressures in the environment of the system, an attempt at symbolic politics.

The interdependence of waste management decisions has been stressed by virtually all students of the problem (e.g., Willrich and Lester 1977: 24). The prime example of this is the question of spent-fuel reprocessing, which would separate valuable plutonium and allow flexibility in the production of a waste form, but presents the potential for diversion to weapons uses. President Carter's decision to defer all commercial reprocessing in the U.S. shortly after taking office had an immediate and direct effect on the nuclear waste system. Technically, the implications were that spent fuel, a relatively unresearched material, was now a waste form for disposal.[3] Innovation was required to prepare spent fuel for permanent disposal and determine its properties in the repository setting. The nuclear industry was affected by the decision as well--in practical, operational terms. Spent fuel, continued to collect in storage ponds, and redesigned racks would only postpone the day on which available space would be exhausted. Predictably, Away From Reactor (AFR) racks were proposed and they met with the same criticism as the earlier RSSFs.

The most important political event in the nuclear waste field prior to the passage of the Nuclear Waste Policy Act was the formation of the Interagency Review Group, comprising representatives of fourteen federal agencies appointed by President Carter. Indeed, its report is the most inclusive effort in the history of waste management to reach agreement on major policy issues. Significantly, six major studies of waste disposal had been released in the two years immediately preceding this effort, including reports by the Department of Energy, the Environmental Protection Agency (EPA), the U.S. Geological Survey (USGS), the American Physical Society (APS), the National Academy of Sciences, and the Government Accounting Office. If there was any consensus which emerged from this coincidence of advisory councils, it was that some of the significant technical and scientific questions pertinent to disposal of high-level waste had not been answered, in particular the suitability of various rock types and the interaction of these media with various waste forms.

The outcomes of the IRG study were varied. Perhaps the most widely announced conclusion was that "the majority of informed technical opinion holds that the capability now exists to characterize and evaluate media in a number of geologic environments for possible use as repositories" (IRG, 1979: 3). That is, the committee did not endorse the adequacy of knowledge to actually *select* sites for disposal. Recommendations included: (1) the adoption of a "systems" approach in which various components of the disposal system were to form a series of redundant and independent barriers to dispersal, (2) needed areas of R&D, (3) a site selection process, (4) the broadening of regulatory R&D, and (5) a process of consultation with state governments in site selection. There was no recommendation concerning the *dissensual* issue of whether nuclear power should continue to grow in the absence of a solution to the waste problem.

Consensus and Conflict

The debates within the IRG, the increased role of environmental groups and critical scientists, as well as the sensitivity of government officials to nontechnical factors are all a shift from the earlier period of neglect, sporadic protest, and "impatience with questions from outside the club" (Bupp, 1979: 123) which formerly characterized the nuclear waste system. But with the increasing *importance* of the system, owing to its locus at the intersection of such key political issues as environmental protection, energy, nonproliferation, and nuclear power, the number of actors and level of conflict have increased. One informant called the area a "battlefield," suggesting there was a "state of paranoia" in the waste management community. Another characterized it as "organized disarray." Many felt the secrecy which surrounded nuclear research during WWII continues to produce public anxiety towards waste management decisions. The opinion is still commonly voiced that nuclear power advocates--including scientists and engineers affiliated with the AEC throughout much of their careers--form a tightly integrated group which attempts to control or suppress damaging information and cannot be relied on to provide impartial technical accounts:

> The national laboratories [have] competent but "gung ho" types, company men. Otherwise they don't last long. [They're] like commando groups, developing solidarity and a sense of invincibility.

Further, interagency conflict over the current program reinforces perceptions of defective managerial coordination (Kasperson, 1980: 137-38).

At the same time, informants--though by no means all--stressed that waste disposal was a problem which could be "solved" from a technical point of view. In the words of one DOE program manager, "there are no show stoppers." A prominent materials scientist claimed, "it's now taken for granted that we can make a packaged waste form with no measurable release of radionuclides for any period of time you care to specify." Another, investigating the release of radionuclides under simulated conditions, said:

> **Most people want to get problems solved and leave waste management alone--[it's been] beat to death....Scientists feel that the nuclear waste problem is solved and if the political crap would stop a safe repository could be built....Politicization is the greatest impediment to resolution of the problem.**

Technical feasibility is even admitted by many critics, such as the Union of Concerned Scientists: "It is our judgment that technical problems can be largely overcome by investigations leading to judicious choice of disposal medium and site selection..., waste packaging and emplacement, and repository design and that none of these matters represents a fundamental technical obstacle" (Lipschutz, 1980: 69). Respondents in the present study tended by and large to express a similar sentiment. Eighty percent agreed somewhat or strongly with the statement "Scientific/technical knowledge at present is sufficient for radioactive waste disposal."

Two points seem central to the conflict over nuclear waste disposal and hence the conditions under which technical actors must operate in producing innovations which can be implemented. First, as in many technical systems, there is a serious gap between "informed technical opinion" and the views of the public on waste issues[4] (Brooks, 1976: 69). If there are solutions to the problem, they have not been implemented and this, more than assurances of their existence, is the most salient and convincing fact in opinion formation. The public is sensitive to outcomes, not potentials, and waste still sits in cooling ponds and storage tanks. Second, the burden of proof for radioactive waste disposal has shifted from opponents to proponents of nuclear power. In many ways the debate over the adequacy of technical knowledge and system progress is a debate over the meaning of the "demonstration of safe disposal." One informant, high on the DOE organization chart, spoke of the ambiguity of the phrase:

> The only way is to come back in 200,000 years and see if the radionuclides migrate. So we don't mean that. And a demonstration in one site doesn't mean we've resolved the technical problems with another.

"Demonstration" rhetoric has been used as a flexible resource by both sides of the debate. Opponents of nuclear power emphasize that "there is not yet a demonstrated means for ensuring the safe, long-term isolation of these wastes" (Lipschutz, 1980: 1). It is often claimed that there is a conflict of interest within DOE between the promotion of nuclear energy and the disposal of nuclear wastes, leading to pressure to implement a program, to push for a quick solution, and to remove the wastes from view and controversy. Asked whether government or industry was responsible for this pressure, representatives from *both* government and public-interest groups said they could not separate the two.

On the other hand, scientists and system managers tend to employ the following kind of formulation: "The analyses performed to date give no indication that a mined geologic disposal system cannot isolate radioactive wastes safely" (Klingsberg and Duguid, 1980). Each side attempts to place the burden of proof on the other, much as litigants who are in an ambiguous situation in court. External conditions, including the nuclear power controversy in general, seem to account for the success of such attempts better than any inherent merits of the arguments themselves. Prior to the 1970s, the greater credibility of government actors and their monopoly of technical expertise allowed them to convince relevant publics of the adequacy of management efforts and the scientific illiteracy of opponents. Subsequently, the emergence of technical experts emphasizing uncertainties, the importance of nuclear waste to the debate over nuclear power, and the sheer passage of time enabled opponents to return the burden of proof for safe disposal technology to the government and produced an increasingly organized and complex technical system.

Innovation in the Nuclear Waste System

In the first chapter the diversity of technical disciplines and specialties accessed by technical systems was discussed. Depending on the nature of the technical problems and the selection of preferred and secondary technical alternatives, different bodies of expertise will be required. Shifts in technical alternatives, the relative importance of specific disciplines at certain points, and the "ripeness" of fields should explain the degree of emphasis and support received by those areas. This section examines the technical problem areas involved in radioactive waste research.

Geologic Media

The emphasis on disposal of a packaged waste form in a geologic medium as the preferred technical alternative enabled the earth and materials science communities to press resource claims in nuclear waste. Of course, nuclear science, including physics and chemistry, is a core discipline involved in every phase of management. Nonetheless, much is known about the basic properties of the radioactive elements and the composition of waste. As the system began to grow and resources became abundant, other scientific fields were recruited and, for their part, sought entry into the system. It would be erroneous to characterize the process as simple advocacy or need.[5] A reciprocal interaction based on the perception of opportunities for resources and information/ legitimacy requirements accounts for the inclusion of new elements.

An example of this process is provided by the recent emphases of the IRG report: "Previously, very few earth scientists have been involved in either program management or scientific R and D for what is now recognized as a problem whose resolution will clearly require an unprecedented extension of capabilities in rock mechanics, geochemistry, hydrogeology, and long-term predictions of seismicity, volcanism, and climate" (1979: 3). The U.S. Geological Survey has been the principal organizational actor promoting the interests of the earth sciences. The scientific debate over the adequacy of the technical basis for geologic disposal has often been cast as a conflict between bureaucracies, the USGS and the Department of Energy.

During the mid-1970s several individuals in the USGS became concerned about possible deficiencies in salt as a disposal medium and drafted a proposal for a series of projects to study these issues. Dissatisfied with the piecemeal response, they published Circular 779, a highly critical evaluation of the state of geologic knowledge with respect to waste disposal. Further, a direct appropriation of research funds was sought from the Office of Management and Budget (OMB). Eventually, the USGS came to provide staff support for the IRG and has been largely successful in proposing technical programs giving geologists a larger role in the waste system. The Earth Science Technical Plan, jointly prepared by DOE and USGS is now a major component of national planning. Contracts listed in support of earth science questions amount to approximately one-half of the budget for the terminal storage program. We may speculate that had other alternatives been preferred during this period of growth, such as space disposal or transmutation, the probability of successful claims by the USGS would have been significantly reduced.

Specific areas of nuclear waste research fall primarily under the headings of waste form research, earth science research, and the interaction of various system components. Given the concept of a mined geologic repository, that is, radioactive waste sealed in an underground rock formation, only exposure of the rock mass or dissolution and transport of the waste by groundwater will cause the waste to reach the biosphere. The former concern is addressed by placing the repository at sufficient depth and by siting criteria which minimize the probability of intrusion through a search for natural resources. The scientific/technological effort is aimed toward understanding and preventing groundwater transport of radionuclides to the biosphere. This problem involves the amount and rate of supply of radionuclides to the groundwater, the length of the pathways, the volume and rate of groundwater movement, and the character of the pathways (i.e., retardation via absorption). Cyrus Klingsberg and James Duguid, in their review of the status of technology, conclude: "Because complete confidence cannot be placed in either the insolubility of the waste form or in geochemical retardation, current research and development is directed toward obtaining a better understanding of the factors that influence groundwater transport" (1980: 16). The notions of multiple barriers and the "systems approach" mentioned earlier are implied by this statement. Performance assessment and risk analysis of the repository system are key elements of the current R&D effort. The evaluation of the effectiveness of isolation in terms of adverse effects on humans is made using failure mode analysis, probability estimation, and consequence assessment. Mathematical models are a primary tool, together with short-term experimental data.

The recent focus on geologic research has meant a broadening of studies of various rock types and candidate sites. The relative preponderance of studies of salt has shifted (Carter, 1978; Gonzales, 1982). Advantages of salt include its extensiveness in the continental U.S. and developed mining technology. Unfractured salt is one of the more impermeable rock types and dissipates heat well. Under pressure it is fluid, and hence self-sealing. This property eliminates fractures, the most likely paths of contact with groundwater. Since salt is highly soluble, salt beds are clearly isolated from circulating groundwater. Finally, the structural strength of salt is high as well as its ability to withstand heat and radiation effects (Ford Foundation, 1977: 256).

The disadvantages of salt have led to a more extensive analysis of other geologic formations. The presence of part of an original salt formation gives no indication of how much has been dissolved and carried away. More importantly, though salt is largely dry, water is present (up to .5%) in fluid inclusions, which could migrate towards the canister when heated (Cohen, 1977; EPA, 1978). Too, elevated temperatures can cause thermal expansion of salt and possible canister movement (though the direction of movement is not clear). Finally, other materials within salt formations are potential problems. Therefore, although it is indeed the "best understood" of rocks (and still, according to some informants, the likeliest candidate), granite and basalt began to receive increased emphasis at the time of the study.

A principal difference between salt and these hard, nonflowing media is the existence of fractures in the latter due to contraction during cooling. Groundwater may pass through these fractures and eventually reach the waste. The tradeoff is their ability to adsorb most radioactive elements, a property which is not present in salt. Should groundwater reach the waste and leach the radionuclides into solution, adsorption along the pathways may reduce the hazard (Kerr, 1979b).

Granite is a crystalline, igneous rock, with advantageous properties of high strength, structural and chemical stability, and low porosity in addition to low water content and valuable sorptive properties. Near the surface much of the water is held in fractures, partly lined with secondary minerals. This implies that groundwater flow will depend on this network of fractures rather than the pore spaces of the crystalline structure. It is generally recognized that modeling of flow through fractured media has recently advanced, but remains a difficult technical problem. At greater depths, stresses reduce the permeability of granite, closing the fractures.

A primary reason for the study of basalt is its proximity to the Hanford and Idaho storage facilities (EPA, 1978; 22). Like granite, basalt has high strength and low porosity, permeability, and thermal expansion. Potential difficulties are the existence of zones of rubble and water-filled sediments composed of altered basalt and other minerals.

While R&D on these disposal media and their interactions with waste forms is important, attention has shifted in the past five years to site-specific studies. The natural variability in sites within a given medium is often greater than that among different media. As one program manager commented: "technical people don't think in generic terms, only in terms of sites and specific problems...nontechnical people muddy the issues by talking generically." Sites are selected based on physical, chemical, and hydrogeologic criteria, tectonic activity, distance from other structures, and possible mineral deposits. The geographical scope is narrowed from regional surveys to area studies and specific locations.

Materials and Barriers

Although the policy decision to defer reprocessing was in effect during the Carter administration and stimulated interest in spent fuel as a waste form, reprocessed waste offers much greater scope for technical innovation. Because its form and composition can be manipulated, materials scientists have viewed waste form research as a felicitous congruence of their professional concerns and technical system requirements. Although there was no bureaucratic sponsor with the interorganizational position of the USGS, the Materials Research Society has begun sponsoring an annual conference on the scientific basis for nuclear waste management which focuses on materials research.

Professional *dissensus* may slow the developmental schedule for the administrative bureaucracy, but it does provide a strong justification for additional resources and stimulates system growth. As in the earth science case, disagreements within a system subfield led to the expansion of the area and allocation of additional resources. Of course competing scientists do not disagree about the fundamental importance of their field. For materials scientists, it was *geologic* isolation which had received undue emphasis: "History has shown that technically this strategy has proved unacceptable. In the last two years the role of the solid waste form and hence the entire radionuclide immobilization subsystem has assumed a much more important role" (Roy, 1979: 1).

Within this area, the debate has centered over the merits of glass and ceramic waste forms. In 1977, borosilicate glass was selected as the reference material and was already in routine production in France. Though ceramics were considered, it was believed that "current information does not indicate any significant advantage for crystallized ceramics" (ERDA, 1977: iii). Then, a series of experiments at Pennsylvania State University in 1978 indicated that at high temperatures and pressures borosilicate glass devitrified (McCarthy, et al., 1978). Since these conditions resemble those likely to obtain in a repository, the search for new thermodynamically stable materials increased in priority (Kerr, 1979). It was argued that ceramics, nonmetallic crystalline materials with an ordered atomic structure, offered an improvement over glass. They could be tailored to fit a particular waste composition and physically trap certain radioactive atoms. However, some scientists interviewed felt that the claims for crystalline ceramics were premature, a position supported by the Hench Report (1979). In an episode shortly preceding this study, a report by the NAS concluding that glass was an inferior waste form was withheld from publication due to internal disagreements (Carter, 1979). Clearly, the question of an acceptable waste form is governed in part by economic and compatibility considerations, but research in these areas continues to be an active field.

Apart from the waste form itself, engineered barriers are being investigated in laboratory and small-scale conditions. Materials requirements such as corrosion resistance, nuclide sorptive properties, and protection of the waste form are similar for most of the waste package components (DOE, 1980d: 5.22). Filler materials for spent-fuel casks include inert gases such as helium to enhance heat transfer. Canister, overpack, and sleeve materials such as metals, ceramics, carbon, and glass-ceramics are being evaluated. Back-fill materials are examined for their ability to reduce radionuclide migration. Finally, accelerated aging tests are used to determine interactions among components of the waste package/host rock system.

Clearly the problems and technical issues involved in radioactive waste management go beyond those enumerated here. No mention has been made of the extensive work in waste transportation, alternative disposal concepts entailing research in space, ocean sediments, and polar ice packs, or institutional and regulatory questions such as public perceptions of risk. Even the narrowly technical problems discussed here are not exhaustive of the range of rock candidates and waste forms which have been and are currently under investigation at lesser levels of support. What the range of disciplines and diversity of problems do reveal is the need for organization and planning, a topic reserved for the following chapter.

The Development of Photovoltaic Cells

The preceding section described the nuclear waste problem in the United States, emphasizing the historical role of the federal government in management and disposal. That the government should organize a technical system to solve the problem of disposition is not surprising, given both the environmental and defense implications of waste, either of which is sufficient to classify it as a collective good.

Energy for residential and commercial purposes is, in contrast, a private (divisible) good. That is, the benefits provided by their purchase may feasibly be withheld from those who do not share the costs. Photovoltaic cells which produce electricity represent a potentially profitable technology for private-sector investment as the costs of alternative energy sources rise and the costs of photovoltaic arrays drop. The federal government is involved in photovoltaics as well as in nuclear waste, but for quite a different reason. Possibly, demand would eventually stimulate the development of a commercial photovoltaic cell through private investment alone if current market trends continue. However, political interests controlling national energy policy have established independence from foreign energy sources as a principal policy objective. The stimulation of energy alternatives in the private sector has been defined as a means to this end. Traditionally it is justified as correcting a defect of the market mechanism: underinvestment by the private sector where appropriability of returns is low or where large risks are involved. Photovoltaic technology, then, is developed in a technical system where investment is split between public and private sectors. The following section presents an introduction to the technology of solar cells and treats their development in historical context.

Fancy-free Energy?

Solar photovoltaic systems are means of generating electrical power by the direct conversion of sunlight to electricity. They are one of a group of "active" solar technologies, along with wind energy and many heating and cooling systems. In contrast to "passive" solar technologies (e.g., architectural design features), photovoltaics are considered a "high technology" alternative. They are often compared in process and growth with the hugely successful semiconductor industry. The photovoltaic effect, a process which occurs when light photons hit certain sensitive materials and create an electron flow, is produced by means of devices made of semiconductors. These "cells" are electrically connected to form modules and are combined into solar arrays. The array is equipped with a regulator, battery, and other components to supply electricity for immediate use or storage.

It is impossible to discuss the history of photovoltaic cells without rehearsing the litany of advantages which have become so much a part of the solar conservationist culture. Central to this are the types of energy used and generated in photovoltaic conversion devices, for sunlight is both unlimited and free and electricity is the most versatile form of energy for technological applications. This only begins an impressive list of features. In operation (though not necessarily in manufacture and assembly) they are essentially free from environmental impacts such as noise or pollution. They have no moving parts, one reason for their low maintenance costs, stability, and longevity. Most current cells use silicon as their primary material, the most abundant solid element on earth. They are easier to produce than integrated circuits and can take advantage of innovations in the closely related semiconductor industry. The efficiency of photovoltaic systems is largely independent of their size, such that small household generators may be nearly as cheap to operate as utility-owned generators. Since transmission costs are reduced if the system is located near its load, dispersed applications might be preferable and the unobstructed surface area required could in many cases be a rooftop for single household dwellings. Such an account makes it readily apparent why photovoltaics have attracted attention from environmentalist and consumer groups as well as energy policymakers.

Of course the principal disadvantage of solar cells is their cost. Current prices of under $10 per peak watt are still twenty times that of most utility-generated power in the United States (Costello and Rappaport, 1980).[6] Balance-of-system (i.e., noncell) costs are high and not as amenable to reductions through technological innovations. More than one analyst has estimated that photovoltaic systems would not be competitive even if cell costs were reduced to zero. But cell costs themselves are problematic. The current labor-intensive technology leads to high manufacturing costs. Those cell types which are expected to lend themselves to cheaper production processes and materials requirements (discussed below) are characterized by low efficiencies. Pure silicon costs are high ($50-$70 per kilogram) and supply is limited, not least because factories are reluctant to expand capacity given the prospect of cheaper "solar grade" silicon in the future. Although sunlight is free, it is not continuous, restricting the market for photovoltaics in many areas. Storage technology, which could overcome this problem, is not well developed. Further, sunlight has a comparatively low energy density (less than 1,000 watts per square meter on earth), implying significant land areas when considered in conjunction with ordinary conversion efficiencies (Fan, 1978).

Finally, the principal difficulty with decentralized applications is the utter lack of experience by utilities, insurers, homeowners, and others. Linking consumer power supplies with the utility grid (an eventual necessity according to most analysts) is likely to encounter numerous problems, both technical and political, and delay widespread adoption. There are, then, considerable grounds for disagreement with the optimistic assessment that "photovoltaics has the potential to supply a large percentage of the world's electric energy demand" (Costello and Rappaport, 1980:338). The debate over appropriate levels of government funding and involvement in photovoltaics is not readily resolved by recourse to a well-defined set of technological features nor a clear-cut comparison with alternative energy supply technologies, but it does illustrate the ease with which physical characteristics of technologies can become part of social accounts.

Solar Cells and Space Flight

The history of photovoltaic research reflects a long period of basic scientific advance preceding the last thirty years of technological development. One historian/scientist claims that this development has not been uniformly exponential, using a "key events" time line instead of the more traditional count of publications. Basic effects and the early device technology did increase exponentially with few gaps until the 1960s, which are characterized by a "relative starvation of events" (Wolf, 1976a: 38). We may attribute this specifically to a shift in the technical criteria relevant to the primary market for photovoltaics. An expansion in the size and complexity of the technical system, as well as a new organizational center followed the transition from space to terrestrial applications of the technology.

The "first" event in the time line leading to the modern development of photovoltaic technology is fairly arbitrary. Many would put the date at 1839 with Antoine César Becquerel's observation that light which fell on a selenium cell produced an electric current and that the amount of current depended on the wavelength of the light--the "discovery" of the photovoltaic effect. Albert Einstein's famous series of papers in 1905 is best known for the special theory of relativity, but he received the Nobel Prize for the one which formulated the photoelectric effect, conceptualizing light as photon particles rather than waves.

The main work leading to the creation of modern photovoltaic cells was performed in the

1950s at Bell Laboratories, RCA, and Wright-Patterson Air Force Base. According to one version, the merger of two teams at Bell Laboratories--one working on selenium cells to generate electricity for rural phone systems, the other developing a device of crystalline silicon--led to a dramatic increase in the efficiency of the best cells through the use of silicon rather than selenium (Flavin, 1982). Work at RCA on an "atomic battery" in which radioactive waste was used to emit beta particles as a power source was delayed, ironically, by the cells' reaction to light (Loferski, 1980). Nonetheless, conversion efficiencies of 6% were achieved there by the mid-1950s, about the time that the cadmium sulfide cell was developed at Wright-Patterson. The persistence of this alternative cell type is partially explained by its relatively early introduction. An ongoing tension between these technical alternatives is still experienced by some researchers.

The early expectations for photovoltaics clearly anticipated an immediate terrestrial market. Primary research objectives were increased efficiency and price reductions. In 1955, two manufacturers took licenses from Western Electric for silicon photovoltaics. Dollar-bill changers and machines decoding punch cards were marketed. An independent photosensor business sold light meters for cameras. For a brief and unrealistic moment the possibility of unlimited electricity from sunlight was heralded. It was not long before oil prices of $2 per barrel and the construction of the first nuclear reactor brought the $600 per watt costs of photovoltaic electricity into perspective. From this standpoint it appeared that research and development on this new energy source would come to an abrupt halt (Wolf, 1976a; Flavin, 1982).

In these pre-Sputnik years, the use of photovoltaics in space had not been considered. A key breakthrough for the industry came with the use of solar cells on Vanguard I for a backup radio transmitter in 1958. Now part of the cultural lore of the field, this event dramatically underscored the longevity of the devices and their potential for space applications. In the absence of a cutoff mechanism, the cells continued to provide power for the transmitter, effecting the uselessness of that radio frequency for other purposes for the next eight years (Maycock and Stirewalt, 1981). After Explorer VI virtually all U.S. satellites used silicon cells, including Skylab, Space Shuttle, and all telecommunications satellites. Initial fears about limits to their power output were successfully put to rest by the installation of up to 1.5-kilowatt arrays on some air force satellites.

This new market stimulated the growth of a space photovoltaic industry. Hoffman (now Applied Solar Energy) and International Rectifier were the first firms to manufacture solar cells. NASA was the primary purchaser of cells during this period. In the early 1960s, Heliotek, RCA, and Texas Instruments entered, creating market saturation and allowing cost reductions from $500 per peak watt in the late 1950s to about $100 per watt in ten years. Initially rapid growth leveled off and production stabilized at 50-70 kilowatts per year by 1968 (Backus, 1976; Wolf, 1974). Three firms had left the market by this time, a factor which may have delayed further price reductions by the presence of a stable demand and an assured market share for Hoffman and Heliotek. R&D during this period was directed toward the relatively specialized requirements of device operation in space. The goal redirection of the late 1950s had had a lasting effect.

New Markets and New Organizations

It was not until the energy crisis of the 1970s that attention returned to the development of a market for terrestrial uses and the associated neccessity for much different kinds of R&D. Rising oil costs increased the price at which alternative energy technologies could be competitive. Dependence on oil production in unstable foreign regimes increased the attractiveness of options which could promise a degree of self-sufficiency. The rhetoric of independence for the consumer played readily to the sentiments of a large and growing environmental and consumer movement. Awareness of the potential of premium markets (relatively high-priced remote applications) to stimulate the industry and promote price reductions seemed to increase. For energy policy planners, this factor was important in creating the sense that government would not continue to be the exclusive client for cell manufacturers.

The role of the private sector (and hence the market) in stimulating a shift in R&D patterns is difficult to estimate, but it is doubtful that it had a major impetus before the increase in government funding for photovoltaics which characterized the 1970s. Once terrestrial use had become a significant policy goal, the technical system grew in size (both research and administrative components) and its determinative influence on the nature of the design criteria was immediate (Jagtenberg, 1975).

With space as the primary market, the cost of cells relative to other means of power generation was a minor consideration. Early systems were installed at up to $1 million per peak kilowatt, a factor of 1,000-2,000 higher than conventionally generated electricity (Madrique, 1979). The primary criteria in space relate to high performance--high electrical output in the smallest possible area, high reliability and longevity, resistance to high-energy particles, vacuum, and large temperature differentials. One instance of innovation in response to these constraints is the effort occasioned by the discovery of the Van Allen Belt in the 1960s (containing high-energy protons), increasing the importance of resistance to radiation damage in cell design. Work at RCA laboratories demonstrated that radiation damage in p-type silicon is less than in n-type, and N-on-P cells were subsequently adopted. A slight decrease in cell efficiency around 1963 was one consequence of the decision, but a lower failure rate was achieved (Wolf, 1974).

Thus, although the current technological program is an outgrowth of the space program, based on technology developed by and for NASA, the cells which existed in the early 1970s were by and large expensive, high-performance devices. A product for the terrestrial market would have quite different characteristics, including weather resistance, but with cost as the dominant constraint. Hence, R&D efforts were redirected once more, this time back to the design criteria which had obtained in the 1950s. During the 1970s, the administrative locus for photovoltaic research moved from NASA to the National Science Foundation (NSF), then to ERDA, and finally the Department of Energy. NSF organized a National Photovoltaics Program in 1972, funding projects at a level of $1-$2 million per year (DOE, 1981). Three new manufacturers entered the market in 1973, a signal that terrestrial applications would be taken seriously by industry (Maycock and Stirewalt, 1981).

At about this time Joseph Lindmayer developed the "violet cell," which at 18% was the highest efficiency hitherto obtained, an achievement made possible through the use of the violet portion of the spectrum. His new company, Solarex, was largely dependent on two NSF grants in its infancy (Lindmayer, 1980). With R&D devoted exclusively to terrestrial applications, the price of cells quickly dropped from $100 per peak watt to around $20 (Morris, 1975). By 1976, the market had clearly shifted, with sales for terrestrial applications exceeding those for space applications by a factor of ten (Costello and Rappaport, 1980). One-third of this market was sales to the federal government.

In 1975, the NSF program team was transferred to the newly formed Energy Research and Development Administration. Federal support for photovoltaics had peaked at $5 million per year during the preceding decade, but had dropped to less than half that amount by 1970. By the time the program was moved it had increased to $8 million (Herwig, 1974).

To this date, the major legislation pertaining to the photovoltaic system is the Solar Photovoltaic Energy Research, Development, and Demonstration Act passed on November 4, 1978 (Public Law 95-590). This seemed to insure steadily increasing levels of federal support in the intermediate term. A plan calling for expenditures of $1.5 billion from 1979 to 1988 was authorized, along with massive production increases (doubling each year to a cumulative 4 million kilowatts by 1988) and a $1-per-watt price goal. Further legislation was passed the same year encouraging federal agencies to buy and install photovoltaic systems

(Federal Photovoltaic Utilization Act). Finally, the Windfall Profits Tax of 1980 allowed tax credits of 40% on the first $10,000 of a system to stimulate homeowner purchases.

Resources for Development

The size of the technical system is dependent on the level of resources derived from each institutional sector with a stake in the output of the system. In the case of photovoltaics, as with many private goods where government defines rapid commercialization as in the national interest, this level depends on three factors: the fit between the technology and the political objectives of federal policy-makers, the degree of scientific and technological uncertainty involved, and the expected market for the end product.

Technology and Politics. The political priorities of the state, whether beating the Soviet Union in the space race or achieving independence from mid-East oil sources, has been a key determinant of federal resource input. During the 1970s the legislation and increased support for solar research of all kinds was in keeping with the emphasis on alternative energy sources and the drive towards energy independence. Budget cuts during the Reagan administration forced the cutback of much photovoltaic research from $150 million in 1980 to $78 million in 1982 and a projected $50 million in 1983 (Flavin, 1982). Much of the project organization (in particular the staff of the Solar Energy Research Institute (SERI), discussed in the next chapter) was dismantled. With an ideological commitment to private-sector development, federal demonstration funds have been terminated along with much developmental work.

Scientific/Technological Uncertainty. Scientific and technical factors are clearly important to conceptions of the degree of both federal and private support warranted. Where development is rapid and uncertainty low, more credence can be given to claims that learning curves and expansion of productive facilities will lead to cost reductions. In the case of photovoltaics, associations with semiconductor technology have tended to produce optimistic growth estimates. Due to their common dependence on single-crystal silicon, the photovoltaic and semiconductor industries developed together for a short time in the 1950s, the latter stimulated by the growing demand for computer technology. Both technologies required the purification and growth of single-crystal silicon and relied on similar manufacturing processes.

With miniaturization, R&D for semiconductors had less relevance for photovoltaics, since the reduced size necessary for computer electronics diminishes the electrical output of cells. Thus, contradictory design criteria led to divergence of the two industries. Nonetheless, they are still considered similar enough that learning curves for the semiconductor industry are used to predict price reductions possible in photovoltaics (Williams and Stephenson, 1979; OTA, 1978). It is frequently claimed that the "knowledge gained by industry during the rapid development of semiconductors in the last two decades could be very useful in accelerating the development of photovoltaics" (Costello and Rappaport, 1980: 351). For most system participants there will be a tendency to emphasize the developmental rather than the "basic research" aspects of the technology to buttress arguments for more resources. Of course, basic researchers are an exception to this generalization.

Expected Market. The market for any new technology is a key determinant of the degree to which private-sector resources will be invested in its development. This market may be federal or commercial. The degree to which it must be immediate or potential is dependent on the mix of firms involved. For large petroleum firms, which subsidize photovoltaic development with profits from other products, a potential market may be enough to warrant continued investment. For independent firms, an immediate market is required to generate revenue.

In photovoltaics it is well understood that the price of systems will be the principal constraint on the demand for the technology. However, the existence of markets at various price levels enabled some firms to begin generating revenues immediately. At present, of the possible applications, only stand-alone systems are economically viable. These are systems, generally at remote sites, not hooked up to utility grids. Examples include remote communication repeaters and receivers, sensing stations, lighting systems, signals, and cathodic protection of remote bridges and pipelines, all of which involve high costs if conventional fuels are used. Small-scale applications include hand calculators, used by the Japanese to perfect manufacturing processes in amorphous silicon, digital watches, clocks, and portable battery charging systems. Finally, there are a few relatively large applications, again remote, where photovoltaic systems are virtually competitive, such as ranger sites, military bases, and isolated villages and residences. It is expected that such uses, particularly in developing countries, will constitute one of the first large markets for the industry. Currently, foreign exports amount to between two-thirds and three-fourths of total production (Dietz and Hawley, 1982).

Prospects and Promises

For large purchases, current prices have fallen to below $10 per peak watt, a decrease of 50% in the past five years (Flavin, 1982). Sales of $150 million in 1982 represent the production of 8,000 kilowatts, or 1,000 times the production levels of the space photovoltaic program (an average rate of increase of over 50% annually for the past five years). Thus, respectable growth in sales, production volume, and applications, together with steady price decreases characterize the industry in the early 1980s. In the view of the Harvard Energy Project, "one can certainly say that photovoltaics thus far constitutes a success story" (Madrique, 1979: 109).

Nonetheless, it must be acknowledged that the prospects for rapid development of photovoltaics, while not poor, are extremely uncertain, not least because of budget cuts implemented by the Reagan administration. The construction of "realistic" estimates of market penetration, costs, production levels, and contribution to the U.S. energy supply is virtually impossible given the uncertainties of any developmental effort, the interdependence of photovoltaics with other energy supply technologies, and unpredictable market factors.

What is clear is the dependence of "accounts," including the rhetoric of estimation, on the interests and social position of system participants. One instance is the extraordinary variation in the estimates of the potential contribution of photovoltaics to the electricity supply in the year 2000 ranging from 1% by a professional scientific association (APS, 1979) to 100% by a solar advocate (Morris, 1975).

Another instance is the use of "vocabularies of innovation," which establish the plausibility of resource claims where dependence on the private sector is high. Just as interests are clearly served in radioactive waste by the statement that there are no *scientific* problems left and the engineering problems are minor, there is a tendency to use these same statements, suitably modified, to instill a sense of confidence and, indeed, inevitability with respect to the development of solar cells. Thus, one may observe accounts such as "no new scientific breakthrough is needed, only a great deal of meticulous engineering to put proven ideas into practice" (Chalmers, 1976), and "price reductions are not dependent on technological breakthroughs" (Morris, 1975). These are characteristic of treatments which advocate government procurement as a primary policy and those which seek to promote popular interest in photovoltaics. For alternative purposes, in this case the American Physical Society's advocacy of more basic research funds, the account is likely to be quite different, stressing the small contribution solar cells are likely to make in the near future: "Until a clear pathway to the photovoltaic future has been established, efforts to stimulate a large-scale, low-cost industry are premature....A long-term and innovative R&D program in photovoltaics is needed" (1979: xiii). Here, as in most uses of scientific ideology, the expressed need for research versus engineering and other directed activities is said to depend on the degree of uncertainty involved. Uncertainty is, to be sure, a useful concept, but one that is difficult to assess and quantify, which adds to its flexibility as an ideological resource. However, while these groups are sometimes at odds on the distribution of resources, they are agreed on the need for general support of the system.

Ideological strategies can also be supported by technological achievements. As federal program managers and solar advocates in both the public and private sectors have been eager to point out, cost reductions in the price of cells have been dramatic. Perhaps more importantly, the price goals set by the Department of Energy have consistently been achieved or even exceeded by actual prices for modules and systems. Of course, such consistency leads to relatively high credibility for the system administrators, in stark contrast to the problems of nuclear waste managers. However, there is some evidence to suggest that the long-term projections have been used as an additional resource in the competition for system support. The Department of Energy's stated goal was to reduce the price of cells to seventy cents per peak watt by 1986, a figure which many estimated would make cells competitive with utility-generated electricity. Privately, many researchers expressed scepticism with this projection. Shortly after resigning from a key position in system administration, one expert termed the figure "a straight lie," one which no one in the field really believed. Following the Reagan administration cuts these estimates became more "realistic" (Flavin, 1982).

In any event, one reason for the relatively low costs of cells to date is the intense competition due to underutilization of capacity in the industry and the sale of systems below production costs by firms attempting to increase their market share. A number of complaints have been filed with the Department of Justice against firms for subsidizing their products with oil-generated revenues (Dietz and Hawley, 1982). This has created artificially low prices and fostered an illusion that solar cells are rapidly reaching competitive levels while supporting system claims for imminent commercial readiness.

Technical Issues in Photovoltaics

Silicon Solar Cells

By way of introduction to current research in the field, a look at the operation of a standard single-crystal silicon homojunction cell is provided. This cell type is generally used to illustrate the operation of basic principles (Maycock and Stirewalt, 1981; Chalmers, 1976; Fan, 1978; Wolf, 1976b). At the most general level, light transmitted in discrete energy packets, or photons, provides impetus to free electrons in the structure of a silicon crystal lattice to produce a current. The important facts to be considered are that, first, displaced electrons must not be lost before doing some work and, second, one photon usually displaces one electron, no matter how much initial energy it has.

Silicon atoms have four valence electrons, that is, four electrons in their outer shells which are available for bonding or conducting electricity. In crystalline (symmetrical or patterned) form each atom shares electrons on four sides to create a stable structure. By a process termed "doping," certain atoms of silicon are replaced by atoms of phosphorus with five outer shell electrons, to form "n-type" silicon, which has an excess of electrons and conducts by negative charge. Another layer is formed by adding boron atoms with three outer electrons, to form "p-type" silicon, conducting by positive charge. Holes tend to accumulate in the n-type silicon due to the presence of a very thin barrier or "depletion zone" of static electricity formed between the two layers during the creation of the cell.[7]

Metallic contact fingers are attached to the top of the cell to pick up these excess electrons, which flow into a wire and back through the baseplate into the p-silicon layer, where they are reabsorbed in holes. The wire may, of course, be attached to an appliance, a pump, or a utility power station. The function of the light photons is to keep this process going by crashing into electrons in the crystal structure and promoting them to an energy level where they can conduct electricity. It is best if the light penetrates to a level near the barrier, so that if a photon is absorbed in the n-layer, the hole it sets free is already close to the p-layer, and if it is absorbed in the p-layer, the free electron has a good chance of crossing the junction. In silicon, the n (top) layer is about ½ micron thick, and the total, a relatively thick 1/100 inch, due to the poor light absorption of silicon.

The theoretical efficiency of a silicon cell is about 25%. Efficiency is determined by taking into account the portion of the solar spectrum at which silicon atoms will absorb photons and the minimum energy required to release one electron. A quarter of the sunlight is lost due to underenergetic photons and a third is lost due to overenergetic ones. Since one photon can only displace one electron, the extra energy is wasted. Due to internal cell losses, reflection of sunlight from the front surface, shading by the contacts which collect the electricity, and cell packing density, the actual efficiency of a typical silicon solar cell is on the order of 15% (laboratory cells). However, device design improvements have enabled increases up to 19% (Perez-Albuerne and Tyan, 1980).

Besides silicon, there are numerous other semiconducting materials which have been subject to research and development, though on a smaller scale. Since the greater the efficiency of the cell, the cheaper the electricity at a given cell cost, a large part of this investment has been in achieving higher efficiencies for new materials which are cheaper than single-crystal silicon. Another strategy involves increasing the power density of the light. Finally, process research is directed toward reducing the cost of silicon (and other) devices.

Materials—Thin Films

One problem with single-crystal silicon is in the manufacturing process, which is time consuming and expensive. Ingots are "grown" through the Czochralski process, a method of drawing molten silicon into an ingot by pulling a seed crystal through it. Wafers are then sliced from this ingot, wasting about 50% of the silicon. Many of the new photovoltaic materials, some still in the research stage, are based on the notion of covering a large amount of exposed surface with a "thin film," which will have a lower efficiency than single-crystal silicon, but will be cheaper to manufacture. Often, a single step is used in production. Polycrystalline silicon, amorphous silicon, cadmium sulfide, and gallium arsenide are among the candidate materials.

The oldest and most thoroughly researched thin film material is cadmium sulfide/copper sulfide. Besides doping of silicon to create n- and p-layers, an intrinsic voltage can be created by the joining of two different semiconductors, the principle behind this cell type. Advantages of copper are its high absorptivity of light relative to silicon, its cost, and an inexpensive production technique. However, its efficiencies are lower and cadmium is a toxic material, supplies of which are limited. Nonetheless, it is the only material besides silicon currently being used in pilot production, its cost dominated by the glass or copper backing on which the copper sulfide is deposited. Recently, higher efficiencies have been achieved with these cells (about 9.1%), partly as a result of $3.2 million in research funds (about fourteen subcontracts) invested by the government in fiscal year 1980 (FY80). The efficiency and degradation of this material is receiving special attention (Leong and Deb, 1981; Kelly, 1978; Fan, 1978).

Gallium arsenide is another material which has higher optical absorption of light than silicon. It can be produced in thin films (stretching the limited availability of gallium) or in single-crystal form for use with concentrators, due to its high heat tolerance. The match between its band gap (the amount of energy required to free an electron) and the solar spectrum is better than with silicon, leading to a theoretical efficiency of 27% (up to 25% has been achieved with concentrators). In recent years higher efficiencies have been achieved in thin-film applications (Perez-Albuerne and Tyan, 1980; Costello and Rappaport, 1980).

Silicon may also be used as a thin film--two of the most innovative current technologies use it as the primary material, but require much less of it than single-crystal cells. If molten silicon is allowed to cool gradually, numerous small crystals are formed instead of a large single crystal. Many "grain boundaries" result, which can serve as centers for the recombination of electrons and holes, resulting in a loss of efficiency. Two methods of reducing this effect are the use of larger crystallites and the treatment of boundary edges with hydrogen. Polycrystalline silicon can be formed in a number of ways. As an ingot, the cutting losses are still high, but it may be drawn from a molten state like a ribbon or deposited as a vapor on a substrate. Polycrystalline silicon has reached the stage of "exploratory development" (see next chapter) and has been funded at a level of twenty-seven contracts and $7 million annually. However, many of the basic effects, such as the mitigating effect of hydrogen, are not well understood (Surek, et al., 1981; Maycock and Stirewalt, 1981).

Amorphous silicon has been used to create still thinner films. Amorphous silicon has no crystal properties at all, the molecules being arranged in a random fashion. Within the photovoltaic community, there is a great deal of excitement over this technology, with the hope that efficiencies can be increased to 10% in the 1990s. Since it has no grain boundaries, these recombination centers are not a problem, and absorption of light is much stronger than in single or polycrystalline silicon. On the other hand, other sorts of recombination sites exist, such as the dangling bonds of the silicon atoms. It has been found that large amounts (25%-50%) of hydrogen can help to solve this problem by linking to the silicon atoms. This phenomenon is still partly unexplained, but has resulted in efficiency increases up to 8.5% (Wagner, 1982). Some calculations indicate that up to 15% might be possible (Fan, 1978). The extensive work in solid-state theory and electronics is less applicable to amorphous, due to its lack of regular structure, and much research of a fundamental nature is being conducted. The FY82 program in amorphous materials was projected at twenty-five subcontracts at a level of $5 million (Stone, et al., 1981).

Processes

Regardless of the efficiency of the material or the quality of the junction, photovoltaic cells must be made inexpensive to manufacture. Each of the processes of purification of quartz to metallurgical- then semiconductor-grade silicon (growing single-crystal silicon ingots, slicing, polishing, and doping), admits of simplification and cost-reducing innovation. Another strategy is to avoid pulling the ingot painstakingly from a melt and simply cast it by pouring it into a crucible and controlling the conditions of cooling. For example, one method under development involves allowing a single crystal to cool on a seed crystal and produces a square ingot, wasting much less material in sawing than the standard cylinder. Another process is already in the production phase. Speed, silicon purity, and energy requirements are expected to be among the criteria which will determine the success of these methods. Since photovoltaic applications simply do not demand the same quality as electronic applications, a means of producing silicon which is *less* pure is one area of innovation.

The slicing step has been eliminated as well through the continuing development of silicon sheet growth methods. The cost per unit cell area is lower with these methods, but so far does not offset the loss of efficiency (APS, 1979). Two of the most innovative processes are edge defined, film-fed growth and web dendritic growth, both of which have been highly publicized outside of the photovoltaic community (Fan, 1978; Smith, 1981). Both require less silicon than ingots and are achieving higher conversion efficiencies. The ribbon process consists in pulling a ribbon of molten silicon through a die, which is designed to shape the ribbon, and winding it onto a drum. Sheets are then cut and doped to produce rectangular cells. The dendritic process involves growing a web between two silicon seed crystals above a melt. Sixteen percent efficiency has been achieved (Maycock and Stirewalt, 1981). Automation and scale-up are expected to reduce production costs. These techniques have been demonstrated in the laboratory, but basic limitations on the growth rate of ribbons and their size may reduce the usefulness of these methods (Wagner, 1982).

Concepts

All of the cell types and processes which have been discussed thus far assume a flat-plate collector is used, fixed in position and tilted at an angle. An important addition is the advent of concentrator systems, which only became possible with terrestrial uses. Since cell costs dominate system costs at present, concentrated sunlight, with its concomitant increase in electrical output, can be used with smaller cells, such that system costs are lower. High-efficiency cells are used, generally made of silicon or gallium arsenide. The concentrated light tends to heat the cells, which lowers their efficiency. This is less of a problem with gallium arsenide, but a coolant is generally designed to lower the temperature of the cells. Properly conditioned, the efficiency of most cell types increases with concentrated light. Further, light can only reach the cell target from a limited number of angles, such that a tracking mechanism is required for most concentrators. Even with the losses of heating and indirect light, 25% efficiency has been achieved (Redfield, 1980). It has been estimated, however, that efficiencies of 30% are required to offset the cost of the mechanical concentrating subsystem (Costello and Rappaport, 1980).

Concentration ratios, or the amount that light is intensified by the lens or mirror, range from 10 to 1,000 times. Designs for concentrators themselves have proliferated. There are simple Winston reflectors which do not require tracking except for slight seasonal adjustments. There are parabolic troughs with higher ratios which focus sunlight on a line of cells and require one-axis tracking. There are more complicated two-axis devices, using the Fresnel lens or paraboloidal dishes (APS, 1979:85-90). Other approaches are to split the solar spectrum and concentrate specific frequencies on those materials for which the conversion efficiencies are highest or, in the tandem cell concept, to stack cells of different materials on top of each other. The latter are relatively undeveloped technologies at this stage.

Finally, there is the much-discussed solar power satellite--large photovoltaic arrays in space, which, unimpeded by Earth's atmosphere, would collect sunlight and beam microwave radiation to stations on Earth for conversion. Feasibility studies have been funded by NASA and DOE. Fears about possible hazards of microwave radiation and the high cost of building the arrays with the space shuttle were until recently disarmed by the sheer magnificence of the idea and charismatic leadership of its major proponent, Peter Glaser. Although some researchers felt it was a case of "NASA looking for something to do," the established expertise of this agency increased the likelihood that some such proposal could gain a constituency. Limited resources and cost estimates over $100 billion have recently led to the probable demise of this option.

In photovoltaics, the number of new materials, processes, and concepts for increasing the efficiency, reducing the cost, and increasing the production volume of cells is already quite large. This discussion has not considered all of these by any means, but has focused on major materials and promising cell types.

Innovation and Activity: A Variable Relationship

Given the diversity of technical specialties described for both nuclear waste and photovoltaics, it is reasonable to guess that they vary in terms of both innovativeness and activity. We turn to the survey data, digressing momentarily to the quantitative mode of analysis, to answer a question which naturally arises at this juncture: what is the relationship between the level of *activity* within a field and its level of *innovation*?[8]

From the standpoint of promoting innovation, the activity of organizing a number of distinct areas of research is essential. This includes both the components (or sets of components) of the system product and the competing technological alternatives. In a context where resources are not unlimited, decisions must be made regarding the relative costs and benefits of channeling funds and people into each of these areas. One factor is the perceived benefit of basic research versus applied research, development, and engineering design. These will in turn depend on the levels of private-sector investment in each of these modes. A second factor is the degree to which one or a few technological alternatives have been selected, reducing the requirements for R&D on other options. A third is the degree of development of a component, with research needs declining as each nears completion. However, one of the most crucial factors in decisions to stimulate an area is the degree to which it is currently producing innovation, traditionally termed the "ripeness" of technological fields, their readiness for exploitation.

It is generally accepted that some areas have greater potential and in fact do produce more innovations than others. Further, the level of activity of a field (number of researchers and research time) is an indicator of the degree to which available labor is being used to promote innovation in an area. In terms of *efficiency*, the optimal use of resources would be to allocate differential shares to those areas which are producing a preponderance of ideas and valuable information. This insures that each research dollar spent goes to R&D where there is a high likelihood of return. Of course, in a general sense the greater the activity in an area, the more innovation one would expect. However, the very notion of "ripeness" implies that one cannot squeeze blood from a stone. Thus, the potential for innovation is a constraint on the degree to which investment will make a difference. When a field is producing many innovations, additional resources are justified until the marginal value of the innovations produced begins to fall below the value of the resources invested.

In an attempt to investigate the relationship between innovation and activity in fields within nuclear waste and photovoltaics a classification scheme was devised based on a review of the literature in each field and unstructured interviews with prominent researchers and managers. This scheme parallels the areas of research in both fields described in this chapter.

As a measure of innovation each respondent was asked to "indicate your assessment of the number of new ideas and applications in each of these R&D areas currently." As a measure of activity they were asked to "indicate whether you, personally, have done any R&D in these areas in the past three years." The respondent was then handed a card with a list of ten (radioactive waste) or nine (photovoltaics) R&D areas. Estimates of the number of new ideas were made on a four-point scale ranging from "none" to "many," while research involvement was a simple dichotomy (yes/no).

If the systems allocate resources in the optimal manner there should be a relationship between the amount of activity in each subfield and the number of innovations produced in the subfield. Exhibits 2.1 and 2.2 show each subfield, ranked by the percentage of respondents in the sample who gave it the highest rating possible ("Many new ideas and applications"). The percentage who have done work in the area is also shown.

Exhibit 2.1 shows that the percent rating each nuclear waste area highly innovative ranged from 12% (interactions between granite and waste) to 34% (research on properties of specific sites). The proportion of the sample having worked in the area ranged from 21% (granite) to 48% (risk analysis and performance assessment). In general, there seems to be a cluster of "innovative" areas which includes ceramics, barriers, groundwater flow, site properties, and risk, all of which are perceived as highly innovative by at least a quarter of the sample. Relatively less innovative are borosilicate glass (the standard waste form), transmutation and reprocessing, and the three candidate media (salt, basalt, and granite). These subfields were rated highly innovative by 12%-19% of the sample.

Photovoltaic subfields are shown in Exhibit 2.2. Ratings of highly innovative range

Exhibit 2.1 Innovativeness and Activity of Subfields in Nuclear Waste*

	% Rating "Many new ideas and applications"	% Having Done Work
Research and Development on:		
Properties of specific sites	34%	37%
Risk analysis, performance assessment of repository or waste form	33	48
Groundwater flow and radionuclide migration in geologic media	31	42
Barriers (backfill, canister, overpack, etc.)	28	38
Ceramic waste forms	26	33
Interactions between salt and waste	19	35
Borosilicate glass	18	32
Interactions between basalt and waste	14	26
Transmutation, partitioning, and reprocessing of waste	14	38
Interactions between granite and waste	12	21

*Results based on 111 to 148 cases.

Exhibit 2.2 Innovativeness and Activity of Subfields in Photovoltaics*

	% Rating "Many new ideas and applications"	% Having Done Work
Research and Development on:		
Amorphous silicon cells	32%	19%
Silicon-sheet growth	32	24
Polycrystalline silicon cells	28	44
Gallium arsenide cells	26	42
Concentrator systems	25	38
Single-crystal silicon cells	24	60
Silicon ingot casting	18	11
Cadmium sulfide/copper sulfide cells	10	24
Solar power satellite	9	21

*Results based on 102 to 140 cases.

from 9% (solar power satellite) to 32% (sheet growth; amorphous silicon), while "activity" ranges from 11% (ingot casting) to 60% (single-crystal silicon). Innovative subfields seem to be predominantly thin films (amorphous silicon, polycrystalline silicon, gallium arsenide) and sheet growth processes, though several fields receive essentially equivalent ratings (differences of a few percentage points are neither statistically nor substantively meaningful). There does seem to be agreement that cadmium sulfide cells and ingot casting are not currently innovative fields. The solar power satellite, which had begun to encounter serious obstacles at the time of the study, received a very low rating.

Exhibits 2.1 and 2.2 show that there is significant variation among fields in both innovativeness and level of activity. The overall ranking of fields confirms our expectations based on qualitative interview materials. Our main interest, however, is in the relationship between these dimensions. Exhibits 2.3 and 2.4 plot the relationship between innovativeness (x-axis) and activity (y-axis) using these same percentages. A comparison of these tables shows a strong positive linear relationship in radioactive waste, but a weak relationship in photovoltaics. Exhibit 2.3 shows that for the nuclear waste system those fields (e.g., research in granite and basalt) with low levels of activity are also those with relatively few new ideas and applications, while those fields with high levels (risk assessment, nuclide migration) are relatively innovative.

In contrast, the photovoltaic subfields in Exhibit 2.4 do not display such a pattern. Indeed, the two most innovative fields (amorphous silicon and sheet growth) are characterized by relatively low levels of activity, while single-crystal silicon, which is only moderately in-

Exhibit 2.3 Relationship between Innovativeness and Activity--Nuclear Waste*

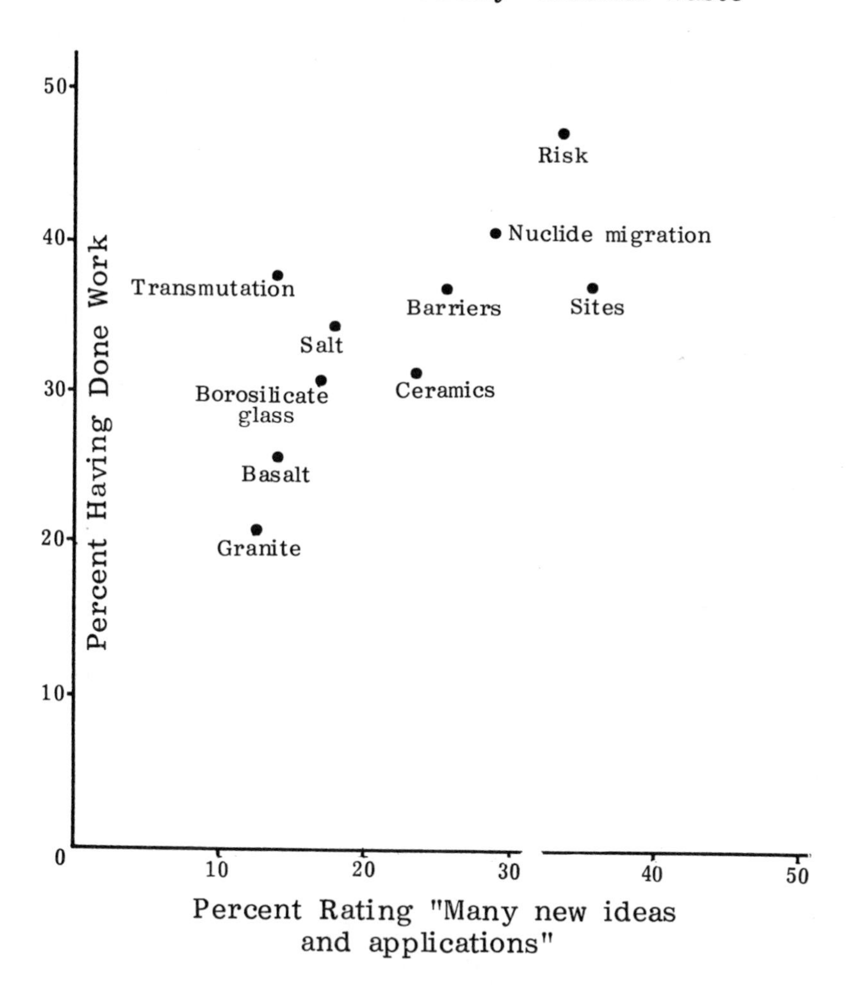

*Correlation coefficient = .754.

Exhibit 2.4 Relationship between Innovativeness and Activity--Photovoltaics*

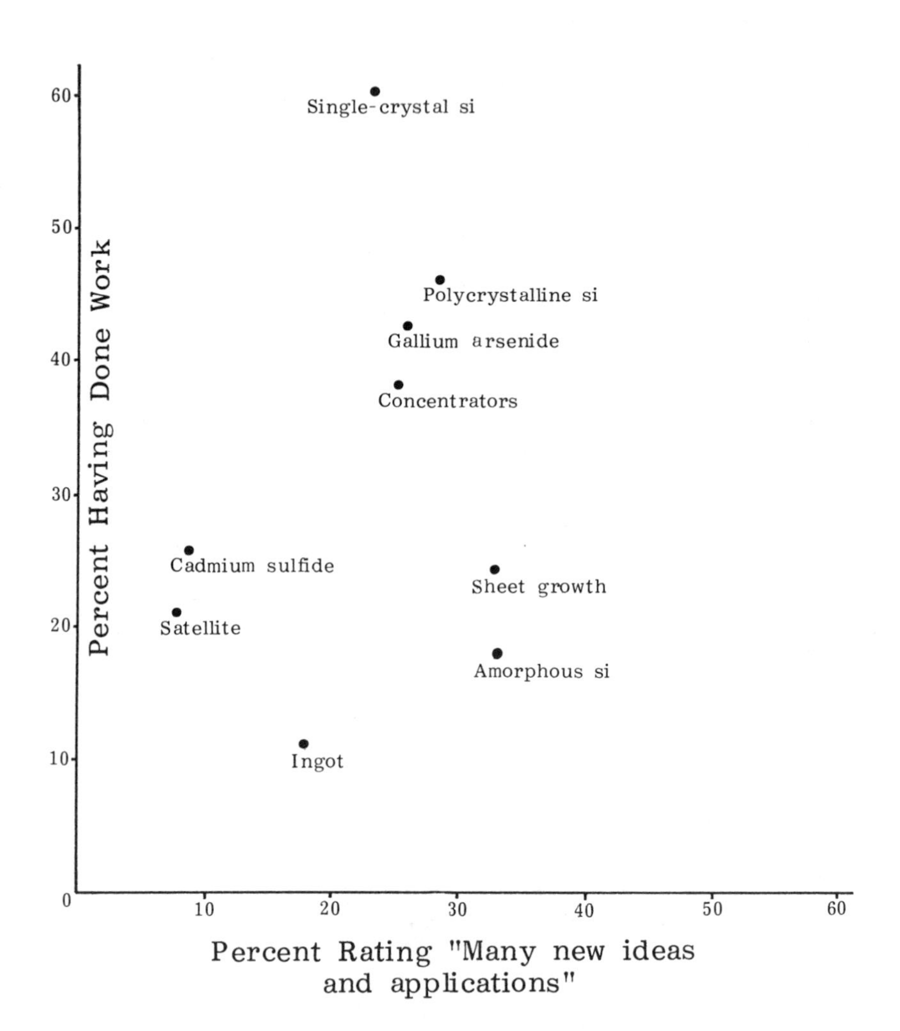

*Correlation coefficient = .284.

novative, has the highest level of activity by far. The two areas which were rated least innovative (solar power satellite and cadmium sulfide cells) were as heavily researched as the two most innovative areas. The visual contrast between the two plots is confirmed by the difference in their correlation coefficients (.75 in radioactive waste and .28 in photovoltaics).

The finding that there is a much closer correspondence between the innovativeness of a subfield and its level of activity in radioactive waste than in photovoltaics suggests that regardless of the overall success of the research efforts in the two fields, the radioactive waste system is more organized in its allocation of resources. That is, the system may make more efficient use of resources by emphasizing those areas of greatest promise. In photovoltaics we find a relatively large proportion of our sample involved in subfields which are producing relatively few innovations, while some relatively innovative areas are undersupported.[8] The suggestion that the photovoltaic system is not operating at peak efficiency is supported by a 1978 report which concluded that technical talent was underutilized (Eckhart, 1978).[9] The question naturally arises whether this difference is due to the operation of the administrative component of the systems or whether some structural feature of the systems constrains the allocation of resources. Indeed, this is the first hint that, whatever the perceived deficiences in radioactive waste management, the degree of managerial control seems to be greater in this system.

Summary

The history of photovoltaics thus far exhibits a number of similarities with that of radioactive waste. Both systems originated in the mid-1950s, though the basic science background (in physics for both systems) had been laid previously. After the initial private optimism over solar cells, the government funded research and development over a prolonged period for both systems and appeared to be the major client for system products. The private sector eventually developed a stake in each system, though for quite different reasons: in photovoltaics, to generate profit from the sales of arrays; in nuclear waste, to eliminate a principal constraint on the growth of the nuclear power industry.

The distinction between photovoltaics, currently redefined as a private good, and nuclear waste disposal, a collective good, is crucial to the differences between the development of the systems, for when the costs of alternative energy technologies increased and the prospect of various commercial markets for solar cells grew nearer, the government was no longer the exclusive client for industry products, and private investment in R&D and production facilities generated a diversity in the sectoral resource base of the system. The sectoral composition, investment, and formal organization of each system will be described in Chapter Three.

Chapter 3

Organization and Participation

The data on innovativeness and activity in subfields presented in the last chapter are far from showing anything conclusive about the nature of the operation of the two systems. The sample was not randomly selected and overrepresents elite members of the two systems. Further, other measures of investment in research might give different results. They do point, however, to possible differences in the operation of systems involving public and private goods, suggesting that there may be factors which differentially affect the allocation of effort and the relationship between effort and innovativeness of subfields.

In our description of the development of these fields the most obvious difference was the relatively greater interest of the private sector in photovoltaics during the past decade. This arises from the profit potential of this product, deriving from its status as a private good. However, the precise role of the various sectors has yet to be established. Universities, private firms, national laboratories, government agencies, public-interest groups, and policymaking bodies all influence the development of these technologies in some way. Since the central administrative component is a primary dimension which distinguishes innovation in technical systems from firm-based innovation, its structure and operation will also be crucial. What is needed to supplement the historical approach is a structural account of the nuclear waste and photovoltaic systems. This chapter presents a schematic model of technical systems and compares nuclear waste and photovoltaics in terms of scale, administration, centralization, and sectoral diversity.

Scale of Nuclear Waste and Photovoltaic Research

The relatively large size of technical systems was exemplified in Chapter One by the Apollo, Manhattan, and Polaris projects. Although we would not expect all systems to be so large--indeed, these seem to have been the largest of all, attracting a proportionate amount of attention--they should be considerably larger than scientific specialties and firm-based R&D projects. Further, in terms of the contrast between public and private technologies represented by nuclear waste and photovoltaics, the dimension of size, so important to many organizational and interorganizational processes, should be relatively constant, in order to clarify the more fundamental processes of administrative influence at issue. If photovoltaics is less efficient in allocating effort to innovation, the possibility that this is the result of large size must be ruled out.

The size of a technical system may be measured in a number of ways, the most obvious being the number of researchers or organizations in an area and the amount of investment. Exhibit 3.1 shows the number of authors (primarily researchers) in each field and the research investment in 1979. A computerized bibliographic search was performed using seven separate data bases to locate all articles,

Exhibit 3.1 Size of Nuclear Waste and Photovoltaic Systems

		RADIOACTIVE WASTE	PHOTOVOLTAICS
ACTORS*	Number of authors (1977-1979)	3,538	2,483
	Number of organizations	294	294
	Number of authors with three or more items	738	678
	Number of organizations (authors with 3+ items)	97	121
FUNDS**	Department of Energy fiscal year 1979 (millions)	$480	$118
	R&D expenditures only (millions)	$158	$102

*Bibliographic search of seven abstracting services (National Technical Information Service, Chemical Abstracts, Engineering Index, INSPEC (physics), Pollution Abstracts, Environmental Abstracts, Energy Information Abstracts) for the years 1977 through 1979.

**Office of Nuclear Waste Management (Department of Energy) Summary Document, March 1980; Photovoltaic Systems Program Summary (Department of Energy), January 1980.

Source: *Social Studies of Science,* Vol. 4, No. 1, Feb. 1984, pp. 63-90.

reports, books, patents, and conference papers which appeared during the three-year period from 1977 to 1979 (inclusive). The outer boundaries of the system of researchers are circumscribed by all individuals who appear as the author or coauthor of at least one item during this period. Although a few persons will be overlooked by this method, it is probably the most comprehensive way of defining the population of researchers in the system. Exhibit 3.1 shows a total of 3,538 authors in nuclear waste and 2,483 in photovoltaics, a difference of about 1,000 authors. These authors are affiliated with 294 distinct organizations in each system.

Although it is clear that these systems are relatively large, it would appear that radioactive waste is considerably larger than photovoltaics. However, this indicator includes many individuals who appear only once as coauthor on a single conference paper and who are thus quite peripheral to the overall innovation process in the system. It is useful to define a "core" and a "periphery" of R&D personnel. The "peripheral" researchers are those who make only one or two brief contributions to an area, perhaps because it relates to an ongoing research program. "Core" researchers remain in the area for a longer (though not necessarily protracted) period of time. In the present study a "core" researcher was defined as one who appeared as an author on three or more items (that is, an average of once a year). Exhibit 3.1 shows that when only core researchers are considered, the two systems are relatively equal in terms of size, with 738 authors in radioactive waste and 678 authors in photovoltaics. Defining core organizations as those which employ at least one core author, we discover there are actually *fewer* organizations in nuclear waste than in photovoltaics (97 versus 121). This indicates that a slightly larger number of core authors in nuclear waste is concentrated in slightly fewer organizations. In contrast the photovoltaic core is more dispersed.

The second indicator of size is the amount of research funds allocated to the system. Department of Energy figures for fiscal year 1979 show overall expenditures of $480 million in radioactive waste and $118 million in photovoltaics. However, the figures for both systems include administrative and operational costs, in addition to myriad nonresearch tasks. This is especially true of the waste system, where handling and temporary storage of the wastes themselves are ongoing activities absorbing a large share of the budget. R&D costs are extremely difficult to separate from overall totals due to the incorporation of such costs with other elements in distinct technology programs (e.g., airborne waste, HLW, LLW, TRU waste, and transportation). However, funds for "terminal isolation" R&D (the National Waste Terminal Storage program) were $158 million, or one-third of the total radioactive waste budget. In photovoltaics, "operations" are essentially federal procurement activities. Subtracting the Federal Photovoltaic Procurement Program from the total leaves $102 million, or 86% of the photovoltaic budget.

It has been hypothesized that in the evolution of a technical system, the proportion of funds devoted to R&D is initially large, declining as funds are devoted to production and operation (Sapolsky, 1972). However, the much larger share of the photovoltaic budget devoted to research does not mean that as a system it is relatively less developed, but rather that radioactive waste management inherently involves mammoth operational costs.[1]

Hence, the proportion of the budget devoted to R&D will always remain relatively small in nuclear waste.

Although these methods underestimate waste R&D by an unknown quantity (perhaps as much as $100 million), private-sector investment in photovoltaic R&D equals or exceeds this figure. We may conclude, then, that although there are small differences in the number of actors and amount of funds in the two systems, they are approximately equal in terms of size. Whatever the differences between them, they are not due to size differentials.

A Simple Model of Technical Systems

As an entity for collective problem solving, a technical system comprises three principal functions and various structures for performing those functions. Given the characterization of a system as a network of actors oriented toward the solution of a set of related technological problems, we may specify knowledge *production* and *distribution* as functions relating to the ends of the system and *administration* as a function relating to means. Although these functions are analytical and may be distinguished in all technical systems, the structures which emerge to perform them vary.

Research is typically carried out in academic institutions, private firms, and national laboratories, all with distinctive sets of organizational goals, values, career paths, and reward structures. Although the distribution of effort will differ among systems, it would be unusual for a large system not to include at least some representation from each of these sectors. The research activities of scientists and engineers in a multiplicity of projects ranging from basic research to engineering design must be coordinated, evaluated, and funded by government program offices and management firms. Once again there is a great deal of variation among agencies as to how this function is handled (Wirt, et al., 1975), but the size and complexity of systems make it imperative that some form of coordination and control structures operate. Since these structures must receive resources from external organizations, they are positioned in an interorganizational network including a number of actors at the federal level (congressional committees, executive agencies, quasi-governmental agencies).

The distributive function involves the utilization of knowledge produced by the research component and can take internal or external forms. Knowledge may be used internally by other system components (research and manufacturing) or it may be disseminated to institutions outside the boundaries of the system itself (other governmental agencies, private firms, other technical systems, the public at large, other countries). The schema in Exhibit 3.2 is purposefully crude, distinguishing only the principal organizational types likely to be found in most technical systems.

Administration

From the standpoint of innovation, the fact of *formal* organization does not distinguish a technical system from a firm. Firm-based innovation is controlled through a staff of professional managers with stable careers in a bureaucratic organizational context. Resource allocation decisions are made in accordance with broad organizational objectives in mind as well as specific product development efforts. Scientists and engineers are assigned to projects, work on problems which arise out of

Exhibit 3.2 Model of an Idealized Technical System

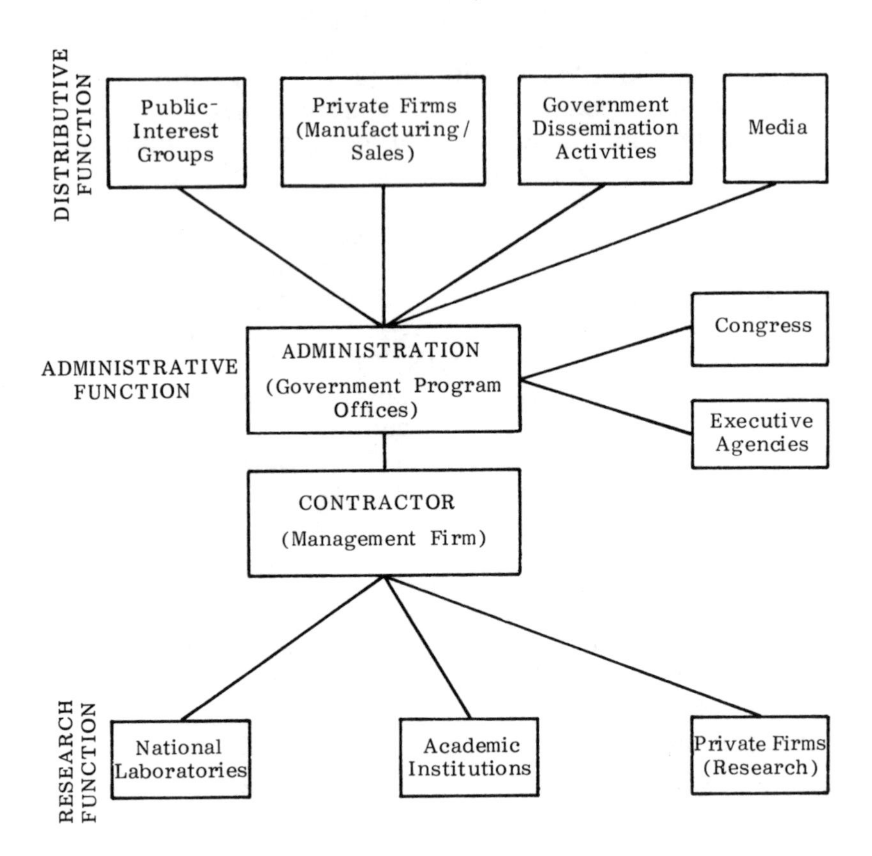

Source: *Social Studies of Science*, Vol. 4, No. 1, Feb. 1984, pp. 63-90.

organizational needs, and write technical reports on their findings to be circulated within the firm. All of these activities take place within technical systems as well. What distinguishes the system is the relatively larger size and complexity of the enterprise and in particular the *autonomy* of component organizations.

What problems does technology development entail for the manager? In the typical case, the private researcher and his/her research manager are employees of the same firm. The manager has a significant degree of control over rewards and career prospects for his/her subordinates. In a technical system, the manager and researcher often belong to different organizations. Interorganizational relationships offer significantly less potential for control, given that the researcher is still embedded in the reward structure of the employing organization. In essence, he/she encounters two reward structures simultaneously. A problem for both managers and students of technical systems is how a collection of autonomous organizational units can be integrated in the knowledge production process. In the vocabulary of interorganizational analysis, how and with what consequences can a loosely coupled system be controlled?

Of course, it is not surprising that a large-scale technological enterprise would require that significant resources be devoted to coordination of the entire effort. "Systems management," "program management," and "weapons systems management" have long been used to describe managerial philosophies and practices designed specifically with control objectives in mind. Reviewing the organization of the U.S. space program in 1965, Fremont Kast and James Rosenzweig wrote: "the systems concept provides the most useful way of thinking about the job of organizing and managing large-scale programs. It provides a framework for visualizing internal and external environmental factors as an integrated whole. It allows recognition of the proper place and function of subsystems" (1965: 286-87).

As a consequence of these concerns the central program office has emerged as a new organizational form designed with the objective of providing a central decision-making and resource-controlling function for the entire system. It should be noted that the presence of "systems" elements as separate administrative structures *within* program offices seems fairly ubiquitous in the field of technical administration. Apart from their practical functions, they serve important symbolic roles, assuring oversight bodies, Congress, and the public that matters are being considered "as a whole," in some sort of integrated fashion.

Like the "project" as against the "functional" or "departmental" organization within single organizations, the central program office is an attempt to focus activities around a specific technological objective rather than performing a set of common activities or a set of activities with a common substantive focus. In general, the larger the project, the more likely it is to have program management rather than functional management. Most central program offices are lodged within agencies of the federal government, but delegate variable degrees of administrative responsibility to external management organizations.[2]

Both nuclear waste and photovoltaic systems are organized and coordinated through the Department of Energy. Each of the organizations in both systems is linked, either directly or indirectly, to the DOE through some form of contractual, consulting, or managerial relationship. Most scientists and engineers in these systems have at least infrequent contact with DOE personnel through site visits or sponsored meetings (Chapter Four). The managerial responsibilities, while highly formalized, are relatively decentralized, making use of national laboratories or private contractors for subsystem developments and integration.

Central Management Agencies

The nuclear waste management program is organized by the Office of Nuclear Waste Management (ONWM) in Germantown, Maryland, near Washington, D.C. Another unit, the Office of Basic Energy Sciences (BES) funds basic research in all energy areas, but developed an emphasis on radioactive waste disposal in 1978 and supports research in chemistry, materials science, and geoscience in cooperation with ONWM. In 1979 these BES expenditures were estimated to reach $184 million over a ten-year period (DOE, 1979b).

The Department of Energy is organized into assistant secretaries for the major energy technologies (e.g., nuclear energy, fossil energy). During the period of study the Office of Nuclear Waste Management reported to the Assistant Secretary of Nuclear Energy.[3] It is staffed by sixty professionals and organized into a projects staff and four divisions, the most important of which are the Division of Waste Isolation and the Division of Waste Products. Each division coordinates, budgets, and evaluates the performance of programs within its area, but projects are actually managed by DOE operations offices. Defense and commercial waste activities are separately budgeted, but in many R&D areas, given the similarity in waste forms and handling technologies, they are planned and managed jointly. Depending on where the first application is likely to be, R&D is funded with commercial or defense budgets. Applied work is often funded by the specific program using the technology (DOE, 1980b).

The Photovoltaic Energy Systems Division is lower in the departmental hierarchy than ONWM, reflecting its status as a unit meant to facilitate private-sector commercialization rather than develop and operate a technology. The division is located under the Office of Solar Applications for Buildings, which in turn is beneath the deputy assistant secretary for solar energy. At the same organizational level are the Division for Active Heating and Cooling and the Passive and Hybrid Division.[4] Like the Office of Nuclear Waste Management, the Photovoltaic Energy Systems Division develops overall policy, budgets, and approves management activities as well as representing the photovoltaics community before Congress and other federal agencies. The division is subdivided into collector research and development, systems development, and market development branches, but remains quite small, with about eight professionals.

Interagency Relations

It is apparent that there is a difference in the relative organizational position of the two principal DOE units, associated with larger staff, administrative budgets, and hierarchical level. Yet the short-term managerial functions of each office (including project management, planning, contracting, and program review), are carried out elsewhere. The explanation for the difference seems to lie in the number and variety of interorganizational relationships which must be maintained in the nuclear waste system. In the previous chapter the turbulent history of waste management was said to involve sporadic conflict with other federal agencies (particularly during the AEC period). The fourteen agencies involved in the IRG are one indication of the complexity of nuclear waste as a policy area.

Regulation of radioactive waste is divided among the DOE, the Nuclear Regulatory Commission (NRC), the Environmental Protection Agency, and the Department of Transportation (DOT). This function depends on the producer of the waste (defense waste is regulated by DOE; commercial high-level waste by the NRC), the phase of management (DOT regulates transportation and handling while DOE and NRC regulate disposal), and the degree of specificity of the regulatory criteria. The EPA issues general guidelines for permissible radiation levels in the environment. For all radioactive waste, general or "umbrella" environmental criteria are set, then specific numerical standards for high-level waste, TRU waste, low-level waste, etc. This work is done by the Office of Radiation Programs in the Radioactive Waste Standards Branch of the EPA.

The NRC licenses sites and repositories for commercial (and eventually defense) waste.[5] General criteria set by the EPA are utilized by the NRC to develop site-specific criteria and licensing standards. The Division of Waste Management is responsible for most of NRC's regulative work, together with the Waste Management Research Branch. DOE and DOT also use EPA criteria for activities which they regulate. Further, DOE responds to NRC standards in developing sites and repositories to be licensed by the latter. One problem which has been of concern to planners is the sequencing of activities by these agencies. The NRC and DOE activities which depend by statute on EPA rules have been underway in the absence of these rules. The DOE is also proceeding without specific licensing standards set by the NRC. Organizational pressures, therefore, may influence the EPA not to set standards which upset these ongoing activities (Greenwood, 1979).

The DOT has authority to regulate the transportation of all hazardous wastes; hence, there is regulatory overlap with the NRC. A memorandum of understanding exists between these agencies for the NRC to develop safety standards for waste packages, which are enforced by the DOT. The level of activity within the DOT is low, with about three persons and no departmental funds involved. DOE funds transportation studies for radioactive waste.

Aside from the problems of regulation, other agencies fund research on aspects of radioactive waste disposal, also requiring monitoring and coordination by the Office of Nuclear Waste Management. Both EPA and NRC sponsor research relevant to the setting of standards, mostly of a "confirmatory" nature. In 1979, the EPA budget for waste management activities was $9 million, while the NRC's FY80 budget for contracted activities (many to DOE laboratories) was approximately $20 million (Greenwood, 1979). However, these research funds are far less than those provided by the Department of Energy, amounting to 5% of the total in 1977 (Kasperson, 1980: 141).

As we saw earlier, the U.S. Geological Survey has been involved for several years in waste management research and has an advisory relationship to both the NRC and the DOE.[6] Much of the R&D program is sponsored by the DOE, although some laboratory and field investigations are performed under separate authority of Congress.[7] Also within the Department of the Interior, the Bureau of Land Management controls federal land and is involved in any activity (such as a waste repository) which utilizes these lands. This intradepartmental relation has been helpful to the USGS in increasing its leverage during interdepartmental conflicts with DOE.

Other federal agencies have peripheral interests in the radioactive waste problem: the Department of State as it intersects with nonproliferation policy, the Executive Office of the President (EOP) agencies as they relate to the environment (Council on Environmental Quality), the science and technology base (Office of Science and Technology Policy), nonproliferation (National Security Council), and domestic political concerns (Domestic Policy Staff). However, these agencies are not investing permanent staff or resources at a significant level and they become involved only at the request of the lead agencies or in policy reviews.

Besides executive and EOP agencies there are relatively long-term congressional interests in radioactive waste, in particular the specialized committees which oversee the operations of the executive agencies, provide authorization, and appropriate funds for nuclear waste programs. While the specific actors change regularly as a result of changes in administrations and congressional turnover, four committees--two in the House and two in the Senate--maintain staff to draft bills and advise congressional committees on waste issues. In the Senate, the Committee on the Environment and Public Works (Subcommittee on Nuclear Regulation) is the oversight committee for the NRC. The Committee on Energy and Natural Resources (Subcommittee on Energy Regulation and Subcommittee on

Energy Research and Development) authorizes DOE funds. In the House of Representatives the Committee on Science and Technology (Subcommittee on Energy Research and Production) is the DOE oversight group. Its counterpart, the Committee on Interior and Insular Affairs (Subcommittee on Energy and the Environment) oversees the NRC. Because of their jurisdictions over NRC and DOE activities and budgets, it may be said that both promotional and regulatory interests are represented in each legislative body.

In contrast to this large and diverse set of federal organizations it is important to note the absence of separate standards and regulatory agencies in photovoltaics. The Photovoltaics Division of the DOE has exclusive stewardship of the federal government program for terrestrial applications and there are no regulatory issues which have as yet warranted separate agency jurisdiction.[8]

As in nuclear waste, there are agencies which are activated for policy reviews and specific issues. These would include those which buy photovoltaic devices as part of the Federal Photovoltaic Utilization Program, and Congress, which has passed legislation dealing with the program. However, after the passage of the Photovoltaic Research, Development, and Demonstration Act in 1978, congressional activity which deals with photovoltaics (aside from yearly program authorizations) has been minimal. Rather than an ongoing problem which requires continual negotiation, monitoring, and legislative response, photovoltaics is a technology which emerged briefly amidst the frantic search for energy alternatives, elicited a small number of congressional actions, and is maintained at a level which does not stimulate continual oversight. This general feeling was expressed by one informant in discussing the photovoltaic budget:

> They're more nuclear inclined than anything. Most [of us] feel they're just covering all the bases....They're really interested in the breeder....In Congress it's regarded as being taken care of for the moment.

The primary interagency relation for the photovoltaic program goes back to the original use of solar cells in space. The initial development of photovoltaic cells for space applications by NASA leads to an underlying tension between NASA and DOE, just as the geological survey has occasionally been involved in domain disputes with DOE in the radioactive waste system. Overt competition is low in both cases, due to clear administrative mandates, but underlying disagreements in approach remain.[9] Two mechanisms have evolved to reduce potential conflict. First, the Jet Propulsion Laboratory (JPL), a NASA installation, has been given the lead role in technology development and engineering, with funds channeled from DOE through NASA to JPL. Second, a coordinating body (the Interagency Advanced Power Group, including DOE, NASA, and the air force) exchanges information on areas of research and funding.[10]

Overall, the structure of the interorganizational network which provides a context for the institutional relations of the central program office is quite different for the nuclear waste and photovoltaic systems. Organizational researchers have hypothesized that the greater the size and diversity of an organization set, the greater the number of specialized boundary personnel required by a focal organization (Evan, 1966; Aldrich, 1979). Hence, we find a relatively large program office in nuclear waste, due in part to the existence of a heterogeneous set of federal actors which monitor its activities, make claims on its staff for information and reports, set requirements for the characteristics of disposal systems, and so forth. A second factor is the relative complexity of the technical core which is overseen by this office.

Decentralization

Government agencies in the United States are characterized by variable reliance on in-house and external technical capabilities. In the military services, the army and navy have relied more on in-house capabilities, while the air force has used outside contractors to a greater extent. The Department of Energy has one of the most extensive sets of national laboratories of any government agency, dating back to the Manhattan Project laboratories of its predecessor, the Atomic Energy Commission. The National Aeronautics and Space Administration developed a vast R&D capability more recently, during the early era of manned space flight and massive investment in space exploration.

The organization of nuclear waste and photovoltaics is constrained by the Department of Energy's policy of decentralized management, itself partly a function of the power and influence of the national laboratories. The form of technical organization is somewhat different in the two systems, but this appears to be conditioned by the nature of the technology under development. In terms of the *kinds* of organizations which must be integrated by the administrative component of the system, nuclear waste and photovoltaics are remarkably similar. Exhibit 3.3 shows the distribution of organizations identified through the bibliographic search by sector. In each

Exhibit 3.3 Sectoral Distribution of Organizations from Bibliographic Search*

Sector	RADIOACTIVE WASTE # of Organizations	RADIOACTIVE WASTE %	PHOTOVOLTAICS # of Organizations	PHOTOVOLTAICS %
Private	135	46	124	42
University	84	29	98	33
National Laboratories	21	7	12	4
Government	40	14	37	13
Miscellaneous	14	5	23	8
	294	101	294	100

*Bibliographic search (see first note, exhibit 3.1)

Source: *Social Studies of Science*, Vol. 4, No. 1, Feb. 1984, pp. 63-90.

system roughly one-third of the organizations are universities, from 4%-7% are national laboratories, and slightly under one-half are private firms. Governmental organizations represent 13%-14% of the organizations in each system (including administrative, policymaking, and other governmental agencies).[11] The integration of private, academic, and national laboratory sectors is a common problem in both systems.

The decentralized management scheme makes use of a complex array of administrative structures under the Department of Energy's central program offices, including field offices, operations offices, and "lead centers." The lead centers are the most important organizations for the technical program, while the operations offices are direct extensions of the DOE located at sites across the country. The formal division of labor provides the following roles: the program office interprets policies, interacts with other federal agencies, develops and defends budgets to Congress and OMB, provides general guidance, and monitors progress at the system level; operations offices and their contractors are responsible both for the day-to-day management of projects (e.g., allocating resources to objectives within cost, timing, and performance constraints), and the management of technology development programs (e.g., preparing technical program plans and integrating the activities of various subcontractors). DOE field activities are also performed by regional offices, whose role is generally limited to grants management exclusive of technical direction. In terms of the levels of organizational function identified by Talcott Parsons (1956), the central program office operates at the institutional level, the operations offices (and their contractors) at the managerial level, and the subcontractors (universities, private firms) at the technical level.

Nuclear Waste Programs

The formal relationships, however, include the contractors who actually operate most of the DOE laboratories, an arrangement which dates back to World War II when federal research capabilities were minimal. The physical facilities are owned by the federal government, but managed by contractors who initiate work proposals, develop technical plans, and execute the technical program through both in-house and subcontracted work. In radioactive waste, many national laboratories are lead centers for subsystems. The long-term waste R&D program is organized by waste category and component, with specific sites designated as lead centers for technology development for each waste category, although R&D is performed at numerous places. Though budget categories are separate for defense and commercial wastes, where the technology is similar the management is combined.

For example, the Savannah River Operations Office was designated in 1979 as the manager for high-level waste technology with technical support from the Atomic Energy Division of the du Pont Company, the operating contractor for Savannah River National Laboratory. Research is performed at many of the waste-producing sites, at universities, and at private firms on the development and characterization of alternative waste forms, processing technology, and immobilization facilities. Du Pont reviews and coordinates all work in the area of waste forms, regardless of the source of the contract and is the key actor for this subsystem.

The most complex example of the decentralized management structure in nuclear waste is the National Waste Terminal Storage program, established by ERDA in 1976 with responsibility for developing a system for permanent isolation of commercial nuclear wastes (Exhibit 3.4). This program is currently in the first, or technology development phase of a four-phase effort which will continue with engineering development, operations, and decommissioning. It is composed of three main elements: the Basalt Waste Isolation Project (BWIP), the Nevada Nuclear Waste Storage Investigations (NNWSI), and the Office of Nuclear Waste Isolation (ONWI). In addition there is a subseabed disposal program managed by Sandia Laboratories. The first two elements are charged with investigating potential repository sites on

Exhibit 3.4 Nuclear Waste Terminal Storage Program Management Schema

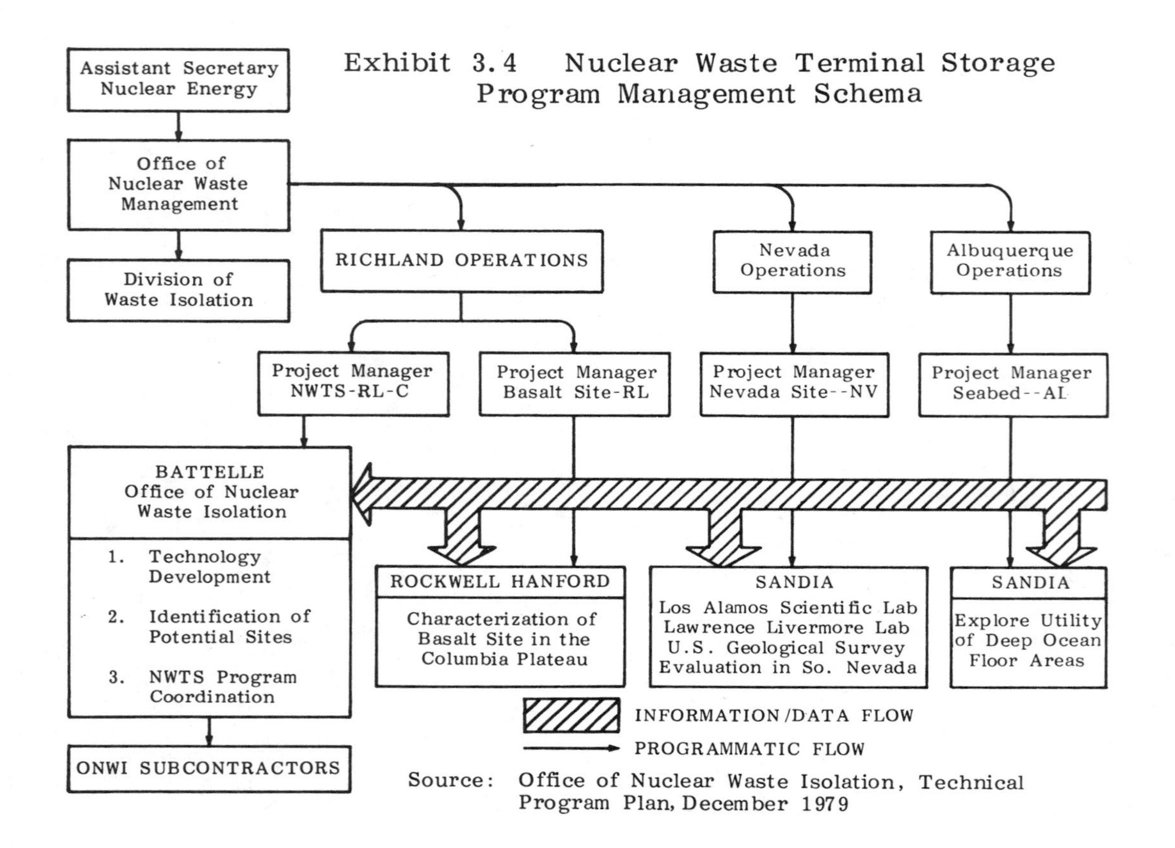

Source: Office of Nuclear Waste Isolation, Technical Program Plan, December 1979

DOE lands. BWIP investigates basalt formations for waste disposal at DOE's Hanford sites in southwestern Washington and conducts work in site evaluation, technology development, facility design, and on-site testing. It is managed by the Richland Operations Office with technical support by Rockwell Hanford Operations. NNWSI performs the same operations at the Nevada test site in geologic media such as granite and tuff. The manager of this program is the Nevada Operations Office, with technical assistance from Sandia Laboratories.

Most researchers in the nuclear waste system perceive the most important organizational actor as the Office of Nuclear Waste Isolation, managed for the DOE by Battelle Columbus Laboratories in Columbus, Ohio. Successor to the Office of Waste Isolation at Oak Ridge, ONWI is responsible for geologic explorations throughout the entire continental United States (all but DOE lands). More importantly, it coordinates all site investigations and manages generic technology (e.g., waste form, package) for the entire NWTS program. This is reflected in the ONWI organization structure as a distinct division (NWTS programs) which coordinates each of the three projects (ONWI itself as well as BWIP and NNWSI) across each of the other system components (waste package, repository, site, and licensing). Another division (technology development and engineering) contracts for generic R&D in geoscience, materials and performance evaluation, field testing, engineering development and design, and systems. ONWI has a staff of about 200, located within Battelle Project Management Division, and is budgeted at about $100 million per year (not including other NWTS elements). Most of this sum is distributed to national laboratories, universities, and private firms in the form of contracts.

Although the management of NWTS is officially the responsibility of NWTS personnel (DOE employees) and is located in the same building as ONWI, ONWI provides the actual technical direction and project management for the entire program. This includes most of the planning functions such as the development of an NWTS technical program plan, an earth science technical plan (in cooperation with the U.S. Geological Survey), a repository licensing plan, and other plans and criteria. The managerial arrangement for ONWI's program integration function was even more cumbersome until October 1980. The lead contractor, ONWI was overseen by DOE's office in Columbus, which reported to its parent organization, the Richland Operations Office (manager of BWIP), which in turn reported to DOE headquarters. Though the chain was temporarily shortened by the elimination of the Richland link, a more recent realignment requires the new NWTS program office to report to the Chicago Operations Office instead of DOE headquarters, once again leaving ONWI a distance of three links away from ONWM in its formal reporting relationships.

Thus, in the nuclear waste system we find a relatively complex administrative structure wherein the organizational units with formal authority have little to do with the management of individual projects or programs and serve as bureaucratic links to the central program office. Informal channels often bypass these intervening formal links and communication takes place directly between actors which are not linked directly according to the formal structure, e.g., ONWI and ONWM. Of course, the existence of such multiple hierarchical levels does not go unnoticed by researchers, who are prone to consider the whole arrangement as a bureaucratic nuisance.

Photovoltaic Programs

DOE laboratories were originally developed as nuclear research centers. With the overall de-emphasis on nuclear research at the time this study was conducted, many laboratories sought to expand their R&D activities into other energy areas to maintain staff and resource levels (Office of Technology Assessment [OTA], 1980). Photovoltaic research was one area where most laboratories could utilize their expertise in physics and capitalize on the emerging interest in solar cells for terrestrial uses, although none of these laboratories is involved in photovoltaics to the same degree as nuclear waste.

Most of the photovoltaic research performed by government takes place at NASA laboratories for the historical reasons outlined in the preceding chapter. They receive funds from NASA's Office of Advanced Science and Technology and from the Department of Energy. Lewis Research Center is the largest center for photovoltaic research with approximately twenty-three persons involved in research, technology development, and testing. Some funds are used for external research, with both grants and contract mechanisms employed (unlike DOE, which uses contracts exclusively). Primary areas of interest to NASA are improvement of cell efficiency, reduction of radiation damage, and high-efficiency cells (e.g., gallium arsenide).

Research and contract management functions are often combined in a single position. Both NASA and USGS informants (in the radioactive waste system) brought out this point to contrast their managerial approach from that of DOE. It is not possible to determine whether separation or combination of roles is more effective, but it clearly serves as an important source of identity and ideological resource in those agencies where the division of labor is undeveloped. In the two cases under review here, the combination of roles in a single status is associated with the use of government-owned-and-operated labs as against contractor-operated labs.

Both nuclear waste and photovoltaic systems are divided into subsystems, wherein responsibility for some subset of researchable problems is delegated to distinct organizational units. Nuclear waste is organized in terms of components, whereas photovoltaics is organized in terms of developmental phases. The nature of the final system output is probably the most important factor here, termed "one-of-a-kind" and "production" technologies by Leonard Sayles and Margaret Chandler (1971). The output of the nuclear waste system will be a small number of repositories, perhaps in different geologic media, each built to order as in the production of AWACs planes or space satellites. The construction of a repository requires the development of a number of distinct components as described in Chapter Two (waste form, geologic setting, waste package). The nature of the waste (high level versus low level) sets different requirements for the type of disposal scheme needed. Finally, the search for and characterization of different geologic media is a task of formidable scope. For these reasons the program organization of the waste system is oriented to the association of lead centers with specific media or waste types.

By contrast, photovoltaic technology will be mass produced with large volumes of solar arrays coming off an automated (or partly automated) assembly line. One or a few private firms will produce all the parts of a cell, perhaps purchasing solar-grade silicon as a raw material. The important point is that the scale of the technology is such that it can be handled by one organization, with the main distinction being between technological options which are close to commercialization and those that are still in the basic research stage.

Since the process from R&D to commercialization involves a number of separate, interrelated steps, the administrative framework distinguishes between program elements based on this continuum and allocates particular technological options to separate organizations depending on their state of development. The four major elements include advanced research and development, technology development, systems development, and tests and applications. Two lead centers report directly to DOE headquarters. The Solar Energy Research Institute in Golden, Colorado, handles advanced research and development, while the Jet Propulsion Laboratory handles the remainder of the program (including federal procurement). This involves distributing parts of its program to other laboratories such as MIT Lincoln Laboratories for residential applications, Brookhaven National Laboratory for environmental studies, NASA-Lewis Research Center for remote stand-alone and international applications, MIT's Energy Laboratory for policy analysis, and Sandia for systems development and concentrators. Only one private firm is involved in a programmatic function (Aerospace Corporation for central station projects).

The Solar Energy Research Institute was created by an act of Congress in 1974 as an organizational center for the promotion of solar technology and a symbol of commitment to increase the visibility of the solar energy alternative (Ethridge, 1980). The Midwest Research Institute was selected to manage SERI in 1977, its first year of operation. The organization grew from 30 to 626 staff members in 2.5 years, has had at least two reorganizations, and experienced significant internal conflicts and external criticism (Williams, 1979; *Science*, 1979). Under the Reagan administration it is experiencing severe cutbacks, as is the photovoltaic program as a whole.

Photovoltaics is one of nine solar energy technologies which have been assigned to SERI for overall management, coordination, and development. During this study it was the only technology with a separate division, performing both in-house and external contract activities. Of the $47 million photovoltaic budget, 8.7% was allocated to internal research (SERI, 1981). External contracting is undertaken in a separate unit differentiated into compound semiconductor, advanced silicon, and analysis programs. Staff size is about thirty-five professionals, of whom most are Ph.D.s. SERI distributes 31.9% of research funds to universities, 12.7% to government labs, 7.2% to not-for-profit organizations, and 48.2% to private firms. This represented a total of 150 subcontracts at 93 locations in FY80. Seventy-three percent of these private funds went to large corporations (SERI, 1981).[12] Notably, the Innovative Concepts program relies *less* on universities (21%) and more on businesses (75%), especially small businesses (32%), than the program as a whole.

Funds for SERI's operations are routed from DOE through its Chicago Operations Office to the Midwest Research Institute and reporting requirements reverse this chain. An on-site DOE office monitors overall administrative performance at SERI, while the Washington office tracks the photovoltaic program in particular. SERI holds semiannual review meetings, annual contract review meetings for particular subject areas, and occasional "focus" meetings, or workshops for researchers on a given topic, not limited to its own contractors. In these respects the management of photovoltaics and nuclear waste is very similar.

Technology development and applications is managed by the Jet Propulsion Laboratory on a budget of approximately $100 million (FY80) and a staff of forty-five people. It is responsible for the program components mentioned above in addition to its own Low Cost Solar Array Project (about 115 people and $32 million per year). A matrix management system, utilized by most of the national labs, leads to flexibility in project assignments but also multiple hierarchical relationships, with each scientist reporting directly to a line manager and a number of project managers. Through its history with NASA space projects, it has developed a reputation for decentralized, effective management.

The division of organizational labor by phases has the advantage of keeping together technologies at the same stages and allowing accurate comparisons of development. The distinct applications of photovoltaics (stand-alone, residential, intermediate load, central station) have been used to divide organizational responsibilities in the tests and applications phase. But advanced R&D, technology development, and systems development are each managed by separate organizations. Thus, as each technology progresses from the research through the development and test stages, the administrative locus shifts from SERI to JPL to Sandia. An important but unknown factor is whether the tendency to develop professional and scientific attachments to particular problems will retard this transfer. This potential could be heightened by the fact that the transfer is interorganizational. The first cell technology to begin this transition is polycrystalline silicon. SERI has awarded $3 million to RCA, Honeywell, and Motorola to improve laboratory-scale fabrication processes. As higher efficiencies are demonstrated in the exploratory development phase, scheduled to end in 1983, the program will be transferred to JPL (SERI, 1980). It is too early to judge the consequences of this arrangement, but it seems to be a somewhat different set of problems for scientific administrators than the more typical tradeoffs between characteristics of an interdependent set of system components.

Planning

A solar cell does not operate in isolation. Cells must be wired together and encapsulated, supported on structures strong enough to withstand the elements, controlled to modify their electrical output, and often attached to batteries. These components, which together constitute the photovoltaic system, must be designed as interdependent parts, an activity which is managed by Sandia Laboratories. Since the development of each photovoltaic technology is to a large extent independent of other technologies, planning in this system takes the form of setting milestones, or dates for the achievement of certain goals for a given application or component technology.

Such milestones include technical feasibility of components, technology readiness of components, system feasibility, system readiness, and commercial readiness. Silicon flat-plate collectors, for example, are commercially ready for certain stand-alone uses, residential systems are beginning to reach system feasibility, and amorphous silicon has not yet achieved technology readiness.

In the radioactive waste system, due to the larger scale and complexity of a repository, the integration of research activities is an even more formidable task. As in the design of the formal organizational structure of the system, however, there are models to fall back on, primarily in the aerospace and defense fields. A large number of program planning and control systems based on "logic networking" or "critical path scheduling concepts," mostly designated with such catchy acronyms as PERT, SCAN, and RAMPS, have been created as means of managing task coordination, information flow, and resource requirements in large-scale systems. Due to the perceived success of such systems, there has been a tendency to use them with modification in all such programs, a mimetic pressure leading to structural similarity. This convergence or "institutional isomorphism" results as much from the uncertainty inherent in managerial activities, the common socialization of managers, and the need to have a sense of control over such multifaceted, interdependent projects as from the increased efficiency achieved by the use of such techniques (DiMaggio and Powell, 1983).

Another function of these practices is the insulation of technical activities from environmental scrutiny. The PERT technique, developed in the navy Polaris program, has been described as an externally oriented strategy to gain autonomy for the program through a reputation for managerial innovativeness, but which had little if any effect on the technical program (Sapolsky, 1972). Similarly, it is not clear what effect the multiplicity of management techniques described in the ONWI Technical Program Plan--including a detailed work breakdown structure, data management system, configuration management system, cost/schedule control systems criteria, and public information program--has on the efficiency of the NWTS program. They are clearly important to the DOE and to critics of the program, given the managerial deficiencies of the AEC waste programs and the need for a strong public image.

It would be a mistake, however, to think of these management practices as purely formal. Much of the communication within the technical system is organized by ONWI and other operating contractors in project review and integration meetings. One example is the configuration management system, as it is used in nuclear waste. This seeks to coordinate the various elements in the isolation system by bringing together groups of individuals to discuss technical work, criteria, and interfaces at different levels of generality. At the programmatic level four interface control boards make recommendations to DOE on funding levels and general decisions on waste isolation, transportation, HLW, and TRU waste. A decision on whether the dimensions of the waste package can be changed would be resolved by the Isolation Interface Control Board. At the intermediate level, interface coordination groups, usually composed of project managers from the NWTS elements, integrate work in the areas of waste package, site, repository, licensing, systems, and quality assurance. Finally, ad hoc interface working groups meet for limited periods on technical issues (e.g., reference repository conditions). They are usually led by ONWI managers and composed of contractors in the NWTS program. Such meetings, in addition to periodic workshops, are an important source of information flow in the system.

Peer Review

A final aspect to the system administration is the operation of outside peer review groups which advise, evaluate, and oversee the NWTS program. It is here, perhaps, that questions of managerial and technical performance arise most clearly. Once again, given the past history of the program, a dominant concern is the prevention of the perception by other agencies and the public that the program is run by the old nuclear establishment in a secretive fashion. For this reason peer review groups cannot be composed of "insiders." They can confer legitimacy on the program only if they are composed of scientists who are perceived as disinterested and independent. Indeed, managerial texts recognize such a function explicitly (Sayles and Chandler, 1971). Since university researchers are viewed in this way to a greater degree than researchers from national laboratories or private firms, we would expect them to be overrepresented in review activities. As of January 1980, there were seven permanent committees for ONWI and BWIP and four temporary committees for NNWSI. Of seventy-two members in all, 54% were university affiliated, with 23% in the next largest group (private firms and independent consultants) (DOE, 1980a). This contrasts markedly with the population of NWTS contractors, in which 50% are private (36/73) and only 26% (19/73) are from academic institutions.[13] Peer review groups are constructed with an external legitimation function in view. The difficulty, as one manager commented, is that: "it's hard to find anyone *external* with a program this big."

Sectoral Diversity

Exhibit 3.3 showed the distribution of organizations in radioactive waste and photovoltaics. These data were derived from a count of organizations with which authors in the bibliographic search were affiliated. It indicates the sectoral *composition* of each system, showing that with respect to the inclusion of absolute numbers of organizations in different institutional arenas, radioactive waste and photovoltaics are quite similar. But one organization can invest more or less resources in a system. The sectoral composition of a system does not indicate the extent of *involvement* of each sector in the system.

A better indicator of sectoral involvement is the number of authors in each sector, shown in Exhibit 3.5. Because there can be more than one author from an organization, such a distribution shows the degree to which the output of the system may be attributed to individuals in each sector. Reports of government-funded research are included here, as well as research which is privately funded if it is published or presented publicly. Although the sectoral distribution of publications is not available, it is likely to give similar results.

Exhibit 3.5 reveals the dominance of the national laboratories in nuclear waste and the private sector in photovoltaics, confirming the results of interviews and historical analysis. National laboratories account for 54% of all

Exhibit 3.5 Sectoral Distribution of Authors from Bibliographic Search*

	RADIOACTIVE WASTE		PHOTOVOLTAICS	
Sector	# of Authors	%	# of Authors	%
Private	637	21	879	43
University	367	12	545	27
National Laboratories	1,631	54	304	15
Government	334	11	263	13
Miscellaneous	27	1	33	2
	2,996	99	2,024	100

*Bibliographic search (see first note, exhibit 3.1)

Source: *Social Studies of Science,* Vol. 4, No. 1, Feb. 1984, pp. 63-90.

authors in radioactive waste (nearly a thousand more authors than the private sector) but only 15% of those in photovoltaics. Private firms account for 43% of authors in photovoltaics but only 21% of those in nuclear waste. These sectoral differences are even more pronounced in the productive core of the nuclear waste system (data not shown). Of those who contributed an average of one item per year, 70% are employed by national laboratories.

Universities

Exhibit 3.5 also shows the greater importance of the academic sector in photovoltaics. University authors represent 27% of the total in this system, as against 12% of the total in

nuclear waste. Although some university scientists are supported in radioactive waste research through federal contracts, their role is limited. The difficulty of attracting university scientists--who are generally perceived by government officials as high technical performers--is often lamented by managers. Further, when asked which are the important university research centers, Pennsylvania State University was often the only mention. There were many more nominations in photovoltaics. The University of Delaware, Brown, Colorado State, Florida State, and Southern Methodist, among others have ongoing, wide-ranging research programs in photovoltaics in contrast to the limited, sporadic, and often isolated research efforts of universities in radioactive waste. Though single-crystal silicon is a well-known material, the newer thin films have not received equivalent attention and many of the basic effects are not well understood. That photovoltaics as a field involves more fundamental research probably accounts for the greater interest of university scientists. In our sample, the average amount of research time spent on *basic* research was 19% in photovoltaics, but only 12% in nuclear waste.

Public-interest Groups

A number of the authors classified as miscellaneous in Exhibit 3.5 are affiliated with public-interest groups. These advocates illustrate the problem of boundary maintenance for the system. In one sense, public-interest organizations are part of the environment of the system, monitoring the affairs of the administrative and research components. In another, they serve a distributive function, disseminating information to the legislative bodies and the public at large. Further, they receive funds for studies of risk and feasibility, occasionally bringing them into the technical core of the system. What is at issue here is the inability of a large-scale technical system to maintain the degree of control over its boundaries which can be achieved by even the largest formal organizations. Public-interest groups are often in an adversarial relationship with actors in the nuclear waste and even, surprisingly, the photovoltaic system.

Environmental organizations which focus on nuclear power issues at the national level generally have one or more individuals who deal with nuclear waste. These range from lobbying organizations (such as Ralph Nader's Critical Mass Energy Project) to groups with strong legal staffs (Natural Resources Defense Council), groups with a strong scientific/ technical focus (Union of Concerned Scientists), and information dissemination bodies (Nuclear Information Resources Service). Few organizations specialize in nuclear waste at the national level, though the Sierra Club Radioactive Waste Campaign and the Southwest Research and Information Center are strong regionally. Numerous local groups have emerged to protest waste shipments and disposal. Most groups are small, even at the national level, such that a single staff member covers a range of issues.

The impact of these organizations, through lobbying, litigation, public organizing, and even technical work, is not proportional to their size. Through the mechanism of challenges to environmental impact statements, licensing, and standard-setting activities, public-interest groups bring the courts into executive branch actions. Their concern with halting nuclear power leads to the promotion of technical uncertainty and an emphasis on credibility issues to generate public opposition to waste disposal activities. Though not common to all groups, some see their roles as impeding all waste disposal plans, in order to maintain the leverage irresolution provides on the nuclear issue generally. Another relatively new variety of organization is exemplified by the Keystone Center for Continuing Education, a "mediation group" which views its role as bringing together actors from all sectors and attempting to negotiate a consensus on nuclear waste issues.

In the photovoltaic system public-interest groups generally act in a supportive fashion with respect to the administrative component. As the last chapter emphasized, the environment of the system is characterized by a relative absence of controversy. Policy disagreements over the approach and rate of federal direction, the likelihood of various levels of contribution to the energy supply, the desirability of centralized applications, and the support of large corporations as against small, garage-based innovators generate some activities on the part of solar activists. For the most part, however, photovoltaics is perceived by environmentalists and the public as an ideal solution to the energy problem. Both solar energy groups (Solar Lobby, Institute for Local Self-Reliance, Friends of the Earth) and antinuclear groups involved in radioactive waste issues (Union of Concerned Scientists, Natural Resources Defense Council) advocate photovoltaics as an energy source to replace oil, natural gas, and coal, as well as nuclear power. Limited resources constrain the level of activity any one organization devotes to photovoltaic issues. Since the R&D work necessary in photovoltaics usually requires significant capital, studies are usually less technical in this field than those in nuclear waste. Advocacy usually takes the form of producing public literature and speaking engagements before Congress and citizens groups.

A key indicator of the difference in the relation between these systems and their environments is the close affinity of the industrial association in photovoltaics (Solar Energy Industries Association) and the public-interest sector. One informant in 1980 suggested that the solar industry was using the Solar Lobby instead of their own lobbyists. This relationship is clearly absent in nuclear waste, where the utility and nuclear associations lack a common interest with the environmental groups.

At the level of the individual researcher the structural relationships between the public-interest sector and the administrative sector (antagonistic in nuclear waste, positive in photovoltaics) are often reversed. One can find a significant level of antinuclear sentiment among university researchers in radioactive waste. Here, personal beliefs about nuclear power may reinforce the common interest of both the environmental and academic sectors in uncertainty. For the former, uncertainty contributes to delays and increasing public opposition to the nuclear power industry. For the university scientist, uncertainty is useful as a strategy to increase funding levels. The publication of results or evaluations which are used to undermine the federal waste program has almost always involved academic researchers.[14]

On the other hand, photovoltaic researchers often see public-interest advocates in a negative light:

> I think Friends of the Earth and the Sierra Club have been harmful to the cause.--photovoltaic program manager and researcher.

Though many are aware that environmentalists help to create a favorable environment for the system and increase the level of federal support for photovoltaics in the long run, there is an undercurrent of "antiadvocacy" culture among scientists. Many seem to feel the promotion of solar energy alternatives creates "performance pressures" and is counterproductive:

> They make pronouncements, raising people's hopes beyond good sense. They tell the public "you'll get sirloin at 25¢ a pound next month" and it doesn't happen....I get letters and calls from concerned housewives who want to cut the power lines and thumb their nose at Texas Power and Light.--private-sector researcher

As a subjective impression, it appears that such sentiments are more often found among older researchers than among younger researchers and those who have entered the field recently.

To this point all the peripheral sectors which make up each system have been discussed. Exhibits 3.6 and 3.7 are graphic representations of the systems, showing the primary organizations and sectors involved. Two of these sectors are depicted in large boxes. The result in Exhibit 3.5 warrants our speaking of each system as comprising a "primary" research sector and two subsidiary research sectors. In radioactive waste, this primary sector is the national laboratory. In photovoltaics, it is the private firm. The reason for this difference, as outlined in the previous chapter, is the collective benefit which derives from waste disposal and the profit potential inherent in terrestrial uses of photovoltaic technology.

Exhibit 3.6 Organizational Actors in the Nuclear Waste System

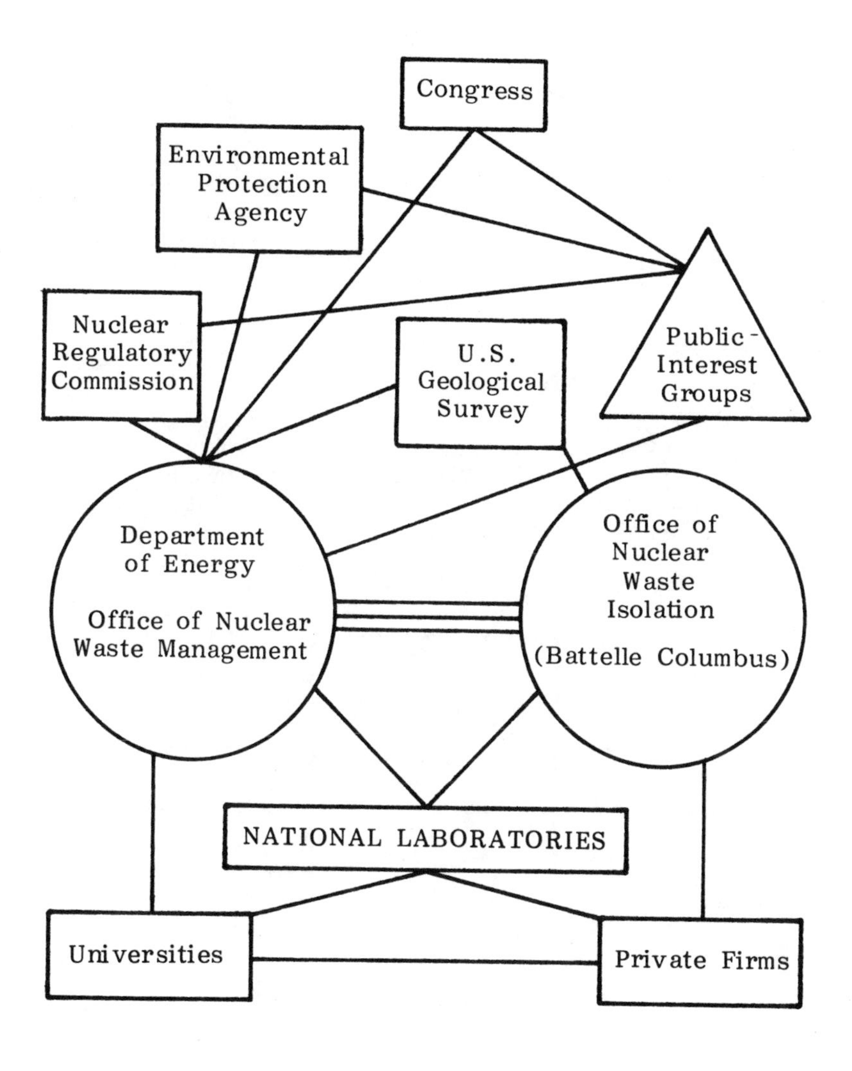

Exhibit 3.7 Organizational Actors in the Photovoltaic System

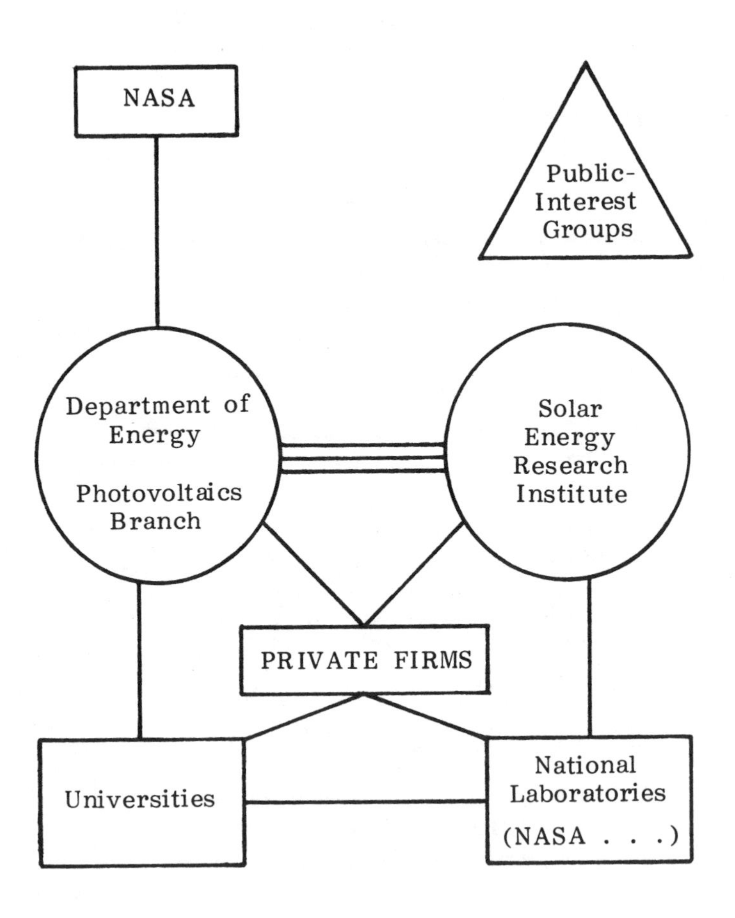

The kind of involvement of each primary research sector is shaped by this factor. The primary research sector in turn has consequences for the nature of the innovation process.

Private-sector Involvement

Of course, it is in the interest of the nuclear industry (the utilities, reactor manufacturers, etc.) to dispose of radioactive waste quickly and convincingly. Solving the problem to the satisfaction of the public (if not the public-interest groups) would reduce the level of controversy and opposition which impedes industry growth. This leads to a small level of involvement in waste R&D, but private investment is negligible. Traditional profit incentives are largely absent (Rowden, 1976). Both the regulatory uncertainty which must be considered in order to manufacture a product to specifications and the long-term nature of repository management reduce industry interest in undertaking R&D except where it is sponsored by the federal government. Informants from the two largest reactor manufacturers expressed the general feeling with regard to radioactive waste research:

> [Name of firm] knows whatever they come up with couldn't be used, so it's not worth the trouble.

> Private enterprise can't make money in nuclear waste. The facilities will be government owned. Thus, there is no incentive there. It's a short term thing. You basically want to solve the problem. [Name of firm] wants to sell more reactors. It's only of indirect benefit to solve the nuclear waste problem.

Industrial associations (the Atomic Industrial Forum, Electric Power Research Institute, American Nuclear Society) sponsor and manage a small amount of R&D with the needs of the industry as a whole in view. Research, consulting, and engineering firms which perform design and engineering studies benefit from the federal R&D program in radioactive waste. Such firms are the primary actors in the private sector.

Owing to the profit potential of photovoltaics as a private good, private-sector involvement in R&D and manufacturing is extensive. Although accurate estimates of the level of private expenditures in the United States are not available, they are thought to be large, amounting to perhaps $200 million in 1982 out of a total $500 million worldwide for both public and private funds (Flavin, 1983). During the middle 1970s government invested a larger share of the total than the private sector, with the two sectors approximately equal at the decade's end (about $100 million per year each).

The increasing investment in photovoltaics by the private sector is largely a function of a change in the ownership structure of the industry: "The photovoltaics industry has rapidly developed from a small industry in 1977 whose typical firm was small, founded and owned by an independent, entrepreneur-scientist...to one which five years later is dominated by subsidiaries of large, diversified global corporations (especially oil based energy corporations) in the U.S. as well as in Europe" (Dietz and Hawley, 1982: 23). Early investment in the form of risk capital and government funds has been supplemented by the immense resources of the oil companies, which have brought up to $100 million per year to the technology (Flavin, 1982). Photovoltaic investments constitute 83% of total oil-firm investment in solar energy according to one estimate (Ethridge, 1980). Unlike the early small firms, they do not depend on current sales and revenue, but can afford to take a long-term perspective, supporting photovoltaics as a new venture in energy, just as some electronics firms (RCA, Motorola) seek a new semiconductor technology.

This change in the ownership structure of the photovoltaic industry has stirred concern that petroleum interests and concentration of production in a few firms will retard innovation and lead to delayed or more costly cells. In spite of the recent influx of private funds, it is charged that these patterns will reduce R&D spending below that which would exist given a competitive market over the long term. The protection of existing capital commitments in energy technologies could lead to suppression of photovoltaics by oil interests. These firms could gain market share by selling under costs, then raise the price of cells to delay their widespread diffusion (Denman and Bossong, 1980; Reece, 1980). This strategy would prove effective only if the market could actually be controlled by the oil firms. It is true that there is a relatively high level of industry concentration at present, with three companies (two owned by oil firms) accounting for 77.3% of the market (Ethridge, 1980: 46). Ten companies which were independent in 1975 have been reduced to one (Maycock and Stirewalt, 1981). However, few analysts expect the current level of concentration to continue, due to the expansion of international markets and the entry of non-U.S. firms (Dietz and Hawley, 1982).

That photovoltaics is in a relatively early stage of development can be seen by the mix of firms. Paul Maycock and Edward Stirewalt have estimated there were over one hundred companies involved in some aspect of the industry in 1979 (1981: 104-7).

Excluding utilities, distributors, and miscellaneous organizations leaves about seventy firms actively involved in production and R&D. Production includes raw materials, components, and systems, as well as original equipment such as watches and calculators. About seventeen firms manufacture systems (two for space applications), while there are twenty-five firms exclusively engaged in R&D. Where the number of nonproducing firms exceeds the number of producing firms, a technology may be characterized as "imminent." This state of affairs in itself increases the risk of firms seeking to develop market share given the potential for technological breakthroughs.

The interaction of administrative and primary research sectors is especially problematic in systems concerned with private goods. Two principal factors lead to this conclusion. First, the balance of market strategies (emphasis by private firms on production and market share) and technology strategies (emphasizing new technologies) is reflected in different advocacy practices. Those firms which pursue market strategy argue for government procurement as a policy to promote cost reductions through an increase in production volume and economies of scale. Those which pursue technology strategies advocate increases in government funding of R&D to supplement firm investments and stimulate innovation. Thus, we should expect system administrators to be subject to pressures from both sides, with much of the policy debate centering on the relative advantages of procurement versus R&D funding.

Second, since organizational restrictions on communication characterize innovation in the private sector, there are greater problems with information flow and coordination than in systems with other primary research sectors. One mechanism which is standard in such systems is to fund organizations without an interest in developing a productive capacity (academic institutions, research firms). Although many organizations, such as subsidiaries of petroleum and semiconductor companies, are participants in the system by virtue of substantial resource commitments, many would withdraw from photovoltaics if government R&D money were withdrawn (Wolf, 1976; Maycock and Stirewalt, 1981; Morris, 1975; Smith, 1981). In profit-oriented systems, a high ratio of research performers to firms with potential productive capacity has important consequences for system effectiveness. It cannot be known empirically what difference the elimination of these firms would make to the onset of widespread utilization of the technology. However, it seems certain that the rate of information flow in the system is greater because of their incorporation than it would be if the research sector were constituted exclusively of firms with productive capacity, owing simply to the relevance of secrecy norms for organizations of the latter type. Research organizations serve an important function for the system in providing more open sources of information, a topic explored in the next chapter.

Another mechanism provided by the administrative sector is the establishment of forums for regular interpersonal communication such as project integration meetings. Such meetings can promote the exchange of information in spite of organizational restrictions on certain types of information flow. Such exchange is recognized by several informants in the photovoltaic field:

> The main source of technical interchange are Project Integration Meetings. Jet Propulsion Lab and Sandia each hold three per year. They're very successful. I'm amazed at the contractors' openness at letting their dirty laundry hang out.... Firms don't have to give out proprietary information, but individuals are extremely open with each other.--laboratory program manager

> Not much is kept secret due to personal contacts.--researcher

> [At these meetings] you don't have to talk about proprietary stuff, only it does make for interesting conversations.--researcher

Over the course of time actors in the system develop cooperative relationships, partly due to their interdependence. Because the sharing of information makes their respective tasks easier it becomes a natural and rewarding activity: the exchange of favors is institutionalized in professional networks. Therefore, at an individual level the conditions are created which promote the systemic objective and defer the organizational one.

Why, then, would a firm participate in a system which undercuts its interests in developing an exclusive technology? As one researcher in photovoltaics put it, "No one will take government money if they can afford it themselves." After all, government contracts require the disclosure of results. Once the industrial firm may have resisted the collectivism represented by central program offices, milestones, planning, and coordination. Contemporary organizations participate because they are constrained to do so. Non-participation reduces the likelihood of innovation and the receipt of information leading to discovery. First, the resource ante required to produce a marketable product has been raised by virtue of the acceleration of innovation produced by state funding. Second, competitors reduce their own R&D costs by taking advantage of system membership. Once a collectivity emerges which is devoted to developing a technology, it is virtually impossible to innovate without participating. Further, the kind of participation required is based on federal funds, which can be used to enhance the research capabilities of an organization as in the case of support for specific projects. Occasionally firms might hold out for special arrangements which reduce their informational contribution in exchange for contract funds.[15] The costs of nonparticipation are almost always greater than the costs of participation.

National Laboratories and the Noncompetitive Environment

The private firm in photovoltaics exists in an interorganizational network characterized by competition and constrained cooperation. The primary research sector in nuclear waste consists in government-owned, contractor-operated national laboratories--Battelle Pacific Northwest Laboratories, Oak Ridge National Laboratories, Sandia Laboratories, and several other major performers. Over 750 national laboratories account for nearly one-third of federally funded research in the United States. Department of Energy laboratories are large multiprogram laboratories operated by contractors--either a private firm, university, or consortium. Such labs should be distinct from "in-house" laboratories (government owned and operated) in their greater autonomy, their more general functions (less closely tied to specific agency missions), and less stringent managerial requirements (they are not subject to Civil Service regulations).

However, in practice they are operated like in-house organizations, with close supervision and control by the parent agency.

Earlier in the chapter the managerial functions of the national laboratories in the radioactive waste program were described. As research performers, their status as privileged organizations affects the process of contracting and innovation within the system.

The degree of competition is usually considered a crucial variable in scientific and technological development. In basic science, individual researchers compete in a race for priority and recognition (Merton, 1973; Collins, 1968). In firm-based innovation, organizations compete to introduce profitable products and to reduce costs through new manufacturing processes (Nelson and Winter, 1977). Within technical systems, the effectiveness of competition for research contracts and decentralization as self-regulating control mechanisms has been described (Sapolsky, 1972). Each of these types of competition is relevant to the innovation process. First, competition for priority in discovery and professional recognition increases the level of individual effort and the quality of research. In systems where technical work is likely to lead to scientific payoffs, it should be greater, although differences among subfields within systems are likely to be as great as those between systems. Second, competition for new products and processes increases the level of private investment in R&D and affects the rate and direction of technical development within the firm. Finally, competition within the research market of a technical system increases the quality of research proposals and output received by the administrative component. These last two types of competition are important for the analysis of differences between nuclear waste and photovoltaics.

For historical and bureaucratic reasons, national laboratories are forbidden by law to compete with private firms for contracts. After World War II, due to the influx of wartime funds and the complete lack of experience of the government with private and academic research funding, the laboratories were considered so well staffed and funded that they possessed an unfair advantage over other organizations. This may not be the case at present, but the legacy of these days remains. They are not allowed to submit proposals in response to a Request For Proposals (RFP). Rather, laboratory staff submit proposals which are reviewed and approved by administrators without undergoing the peer review process which is common when applying to outside agencies for grant support. Such an arrangement has led to criticisms that the quality of work in national laboratories suffers as a result.[16]

Noncompetitive contracts serve an important function for program managers, who respond to a structured incentive to allocate research tasks within scheduling constraints. It was estimated by one manager that it takes up to one year to award a competitive contract, whereas a contract may be awarded to a laboratory in one month. The ponderous array of milestones and performance deadlines reinforces the tendency to use laboratories rather than send out an RFP. As a result many contracts are awarded based on proposals by one laboratory, either unsolicited or prepared by agreement between managers. Besides efficiency, there is a tendency to award contracts to laboratories to maintain their staff and research capabilities: "Agencies with substantial laboratory facilities generally make every effort to use them and keep them healthy, particularly when research funds are scarce. Seldom will an agency fund extramural work unless the maintenance of its own labs is assured" (OTA, 1980: 35). Given such a heavy reliance on laboratories for the research output of the system as a whole, and the absence of contract competition in the funding of laboratory proposals, it is likely that the level of competition is substantially lower in the radioactive waste system than in the photovoltaic system.

We are left with a situation in which two kinds of competition condition the operation of nuclear waste and photovoltaic systems in a complex fashion. Since photovoltaic cells are private goods, the profit potential of the field generates a substantial amount of private-sector investment (in the late 1970s, about one-half of total system resources). These research investments mean that researchers within the system are able to take advantage of a number of alternative funding sources, and are less dependent on government for public funds. Firms which invest are also attempting to innovate products for an emerging but potentially billion-dollar photovoltaic market, leading to competitive relationships and proprietary restrictions on communication. They participate in a system which requires them to disclose a part of their research output because they can reduce their own developmental costs and because the costs of nonparticipation are high.

If the radioactive waste system relied as heavily on the private sector as the photovoltaic system, we would expect contract competition to be high but product competition low. However, as we have seen, high levels of participating national laboratories reduce contract competition as well.[17] Reduced uncertainty in the flow of resources and the absence of a profit orientation allows laboratories to operate in an environment which is relatively insulated from the contingencies which characterize the private sector in photovoltaics. Their greater dependence on government, on the other hand, puts a high premium on relationships with federal program managers to insure continued support.

Summary

In spite of apparent differences radioactive waste and photovoltaics were shown to involve approximately similar numbers of core researchers and federal R&D funds in the late 1970s. Each is administered by a central office in the Department of Energy and a private management firm. One reason for the larger size of the administrative component in nuclear waste is the much larger number of interagency relations which must be maintained at the federal level. Central administrative offices in both systems delegate a large share of the planning and management activities to national laboratories and other lead centers. Nuclear waste, as a "one-of-a-kind" technology, is organized in terms of system components, whereas photovoltaics, as a "production" technology, is organized in terms of phases from basic research to engineering. Complex scheduling and appraisal techniques, project review meetings, and peer review groups are designed and used by administrators in both systems to manage complexity and legitimate their actions.

Sectoral productivity counts showed greater participation of the academic sector in photovoltaics, owing to the more fundamental nature of the research. Public-interest groups, which hold oppositional positions in the nuclear waste system, are involved in some research problems (e.g., risk assessment) and stimulate other kinds of research by traditional organizations. Although they support photovoltaics as an energy alternative, public interest actors are viewed with considerable scepticism by researchers in this system as well.

Productivity counts also showed distinct primary research sectors in each system. In photovoltaics, the large profit potential of a developing terrestrial market encourages private firms to invest their own funds and R&D facilities in addition to relying on government contracts. The cost of participation in a system requiring open disclosure of results is offset by more rapid technology development. In nuclear waste, the relative absence of private interest and the historical role of the national laboratories in nuclear research is augmented by a monopoly of government funds. National laboratories are forbidden to compete for proposals, but while competition is lower, there are gains in efficiency in the placement of contracts. The monopoly of funds by government and the dominance of national laboratories indicates a significantly different interorganizational structure in the nuclear waste system.

Part Two

Communication and Performance

Chapter 4

Communication and Intersectoral Relations

In the remaining chapters data from a national survey of participants in these systems are used to illuminate the following question: "what difference, if any, does the interorganizational structure of these systems make to the innovation process?" By interorganizational structure," the distribution of research funds and involvement across sectors is denoted. By "innovation process" the variety and frequency of communication behavior and its relationship with quality of work at the level of the individual researcher is intended. As a working hypothesis, which will be specified in more detail, it is plausible that the system presents a researcher with opportunities and constraints, and that these vary depending on the type of interorganizational structure which serves as a context. It is important to determine whether similarities in the organization of nuclear waste and photovoltaics are more important than their differences. We have seen that nuclear waste is characterized by a government monopoly of funds and heavy reliance on national laboratories, whereas the private sector is more significant in photovoltaics. Do these broad contextual settings make any difference to the researcher, and the process of constructing new knowledge?

Communication in Technical Systems

The analysis in this chapter is concerned with the nature and frequency of communication in nuclear waste and photovoltaics. Communication is a central dimension in virtually all accounts of scientific and technological development (Mulkay, 1977; Hagstrom, 1965). The transmission of ideas between researchers through formal and informal mechanisms is well established as a necessary condition for much technological advance. Studies of the technological innovation process point to the importance of informal information exchange over formal media such as journals and technical reports due to the efficiency achieved by interpersonal contact (Allen, 1977; Rothwell and Robertson, 1973). One objective of program managers in large-scale enterprises is the stimulation of informal channels through the use of formal mechanisms such as meetings and workshops. Further, a number of studies have alluded to the relationships between researchers in public, private, and academic sectors as a determinant of high levels of technical performance. Currently, intersectoral relations are a central policy concern (Office of Technology Assessment, 1979). Finally, the preceding chapter has described conditions of competition which should lead to differences in the rate of technical assistance among researchers in different sectors. The purpose of this chapter, then, is fourfold:

(1) to describe the overall levels of communication among members of the nuclear waste and photovoltaic systems.

(2) to determine the relationship between formal and informal

communication mechanisms within the systems.

(3) to estimate rates of social linkages between sectors.

(4) to contrast the exchange behavior of the private sector in photovoltaics with that of the other research sectors in both systems.

These objectives require a different approach than the one which has been employed in Chapters Two and Three. Historical records and unstructured interviews are necessary for forming a conception of the larger issues which surround these technical systems, the variation in perceptions of technical and nontechnical issues, and the development of the fields. But they are not as useful as more systematic approaches to the precise description of interaction patterns. Do administrators and program managers maintain personal contact with researchers? Do they have higher rates of contact with some sectors than with others? To what extent do researchers in the field communicate with one another? Do similar patterns exist in systems concerned with collective and private goods? The present chapter provides some answers to these questions.

The Sample and Design of the Survey

Previous studies of technical systems have emphasized the importance of communication among technical and administrative actors but have provided no quantitative estimates of hypothesized patterns (Sayles and Chandler, 1971; Sapolsky, 1972). Recently, advances in the methodology for the study of social networks have made such estimates possible (Beniger, et al., 1979; Berkowitz, 1982). Open-ended nominations and rosters of names may be used as stimuli for directed questions regarding the type and frequency of social relationships between individuals. The present study utilized personal interviews with a roster technique to assess a number of distinct relationships among actors in nuclear waste and photovoltaics, the first use of such a technique with a national sample of individuals connected by a common substantive interest.[1]

The bibliographic search described above generated a population from which the sample of researchers could be drawn. However, the population of actors for the administrative and policy components of the system could not be identified this way. These individuals were identified by a combination of position on organization charts and nominations made during the unstructured interviews. A larger number of nonresearch actors was sampled in nuclear waste to reflect the actual distribution of activities in this field. In order to estimate the extent of communication between actors in different sectors, the sample was stratified by sector, selecting individuals from government and lead centers, national laboratories, universities, private firms, public-interest groups, and key policy organizations. For the research sector, the most prolific individuals within each organization were selected (those who had the largest number of items in the bibliographic search). A total of 297 individuals (152 in radioactive waste, 145 in photovoltaics) representing 97 organizations were interviewed in late 1980. The interviews were conducted in 25 states by a professional interviewing firm and members of the project staff. The final response rate was 92%.[2]

The sample for this study, it should be emphasized, was not randomly drawn and tends to represent an elite within the system. The purposive selection of highly productive individuals was made to increase the likelihood that "boundary-spanning" individuals would be included, reasoning that productive researchers would be more active in the communication network. Administrators and program managers were selected based on organizational position because their authority and coordinating activities derive from their hierarchical status within the administrative component. Finally, public-interest and policy actors were included based on nominations and organizational position where possible. These individuals do not have a formally defined status within the system and must be identified via the convergence of mentions by knowledgeable informants. Except where the level of participation of a particular sector in a system is quite low, this method will generate a relatively elite sample. The results of the statistical analysis cannot be generalized to the system as a whole, which, as we saw earlier, has a large number of "peripheral" actors. However, insofar as the elite is more actively involved and influential in terms of innovation and communication processes, the results of this exploratory analysis may be more significant than could be obtained from a sample which was truly random. Of course, in the absence of resource constraints such a sample would be preferable.

Interpersonal Interaction

Social relationships among actors are not only important to the rate of innovation. They are also the empirical referent of system integration, the extent to which the technical field hangs together as a unified whole. The first question to be answered in an analysis of technical systems is a descriptive one: to what extent do participants (researchers, administrators, etc.) engage in communicative behaviors with other members? That is, does the formal system of relationships specified in Exhibits 3.6 and 3.7 constitute an interactive system as well? For the manager, such a question has a very practical meaning: do the formal structures which have been devised to facilitate the flow of information actually stimulate interaction between individuals? While these cross-sectional data cannot address the causal relationship implied by this issue, we can use them to establish the existence of a relationship between participation in structured system "events" and rates of interpersonal interaction in nuclear waste and photovoltaics.

Communication: Self-report and Network Measures

Exhibit 4.1 presents descriptive information on communication behaviors for the sample as a whole and for each field separately. As mentioned previously, a larger proportion of actors in the nuclear waste sample are involved in policy and administrative activities. When this is taken into account, there are few differences between the fields in overall rates of communication. Indeed, there appear to be relatively high levels of communication in both fields.

The first part of the table deals with communication useful to the respondent's work in the field (nuclear waste or photovoltaics). Over three-quarters of the sample had useful discussions at least monthly with persons outside of their own organization, while almost 90% had useful discussions with persons in their organization. A related item, #5, shows that the typical respondent has contact with eleven persons per week in his own organization who work in the field. Forty-four percent of respondents in both systems found such discussions useful with persons who were *not working in the field* themselves, suggesting that linkages outside the network are important as well and supporting the notion that bridging ties are significant sources of information acquisition in science. The only difference between the systems in terms of communications which are useful to work in the field is in the greater importance of discussions with sales and marketing personnel in photovoltaics. Of course, this finding is not surprising since the products of the system will be sold for profit and contacts with "client" departments may be important sources of information about market needs. Nonetheless, nearly a fifth of the radioactive waste sample found such contacts useful as well.

Several other indicators of communication were used which are not specific to the nuclear waste and photovoltaic fields. However, given the nature of the sample it is likely that much of the activity tapped by these items is communication relevant to the systems. One item indicating rather extensive daily communication of an interorganizational nature is the average of 5.2 phone calls *received* from external callers. The next set of items was designed to assess general levels of communication with specific groups. Exhibit 4.1 shows the percentage of respondents who had frequent dealings (defined as eight or more times per year) by phone, letter, or in person with individuals in other sectors.[3] Individuals in both systems have the highest rates of contact with government agencies, with approximately two-thirds of those in both systems reporting frequent dealings. The

Exhibit 4.1 Informal Communication in Nuclear Waste and Photovoltaics

	Nuclear Waste	Solar Cells	Full Sample
Percent who had discussions in person or over the phone at least monthly during the past year which were useful to their work in the field with persons:			
(1) Working in the field at other organizations	78%	75%	77%
(2) Working in the field at same organization	86%	90%	89%
(3) Outside the field (excluding same organization)	44%	44%	44%
(4) In sales and marketing (including same organization)	18%	38%	28%
(5) Average number of professionals in same organization contacted at least weekly (working in nuclear waste/photovoltaics)	12.1	10.5	11.3
(6) Average number of phone calls per day from persons outside the organization	6.2	4.0	5.2

Percent who had frequent dealings during the past Year: (8-10 times or more) by phone, letter, or in person with individuals from (excluding same organization)			
(7) Government agencies	64%	62%	63%
(8) Business and industry	36%	52%	44%
(9) Universities	28%	34%	31%
(10) National laboratories	53%	38%	46%
NETWORK MEASURES			
Average percent of sample reporting contact with each respondent:			
Frequent contact	4.0%	5.7%	
Infrequent contact	7.0%	10.5%	
Heard of, but no contact	10.5%	14.1%	

results on the remaining items might have been expected from the differential sectoral involvements in each system. We saw that the private sector contributed the largest share of personnel in photovoltaics. Respondents in photovoltaics tend to have more frequent professional dealings with persons in business and academic organizations. Respondents in nuclear waste tend to have more frequent dealings with persons in national laboratories.[4] Given the disproportionate number of researchers in the national laboratory sector this is also to be expected.

Although these measures indicate generally high levels of communication, they share two difficulties with most traditional survey measures of communication. First, they do not distinguish well among actors. The limited number of categories available (four in this case) makes it impossible to differentiate between those respondents who, for example, report "frequent" discussions, though they may be quite different in their behavior. An even greater difficulty is their nonspecificity. The notion of communication implies relationships with specific persons. In terms of reliable measures, it is clear that asking someone how often they have contact with persons at universities is subject to more interpretations and vagaries of memory than asking how often they had contact with specific other individuals. In terms of analysis, for many purposes it may be more valuable to know *whom* one has contact with rather than *that* one has contact. For these reasons a network methodology was used to assess the detailed structure of relationships between individuals in the sample.

The procedure for measuring specific linkages used a roster to elicit responses for a variety of types of social relationships. Our primary interest here is in communication, for which an item measuring frequency of "professional dealings" was used. Each respondent was handed a list of the names representing the sampled individuals in each field (nuclear waste or photovoltaics). The interviewer then asked:

> For each of the individuals listed here, please indicate whether you have *never heard of* the person; have heard of but have had *no professional dealings* within the past three years; have had *infrequent dealings*, that is, between one and three per year; or have had *frequent dealings*--four or more times per year over the past three years.

The principal advantage of network-based measures is that indicators may be derived which do not depend on the perceptions of the respondent alone. The responses of *other* actors in the system may be counted to estimate the number of relationships for any given actor. Thus, we may measure communication "objectively" in the sense that we are relying on a number of independent observers to report social interactions--it is a matter of indifference that they are the participants in these interactions, whose behavior is reciprocally reported by that same actor.

The final percentages in Exhibit 4.1 show the average number of social contacts for respondents in each system. "Other-reported" counts of relationships are used, standardized by the number of names on the roster for each system (168 for radioactive waste, 156 for photovoltaics). In radioactive waste, respondents had frequent or infrequent professional dealings with an average of 11% of the sample. In photovoltaics, contact with an average of 16.2% of the sample was reported.[5]

Percentages for the three categories for this variable show that communication partners within the system may be thought of as a series of concentric circles with the researcher (or administrator) at the center. Individuals with whom the respondent has frequent dealings represent a relatively small set (4% in nuclear waste, 5.7% in photovoltaics). Surrounding this inner core of "strong ties" is a larger number of "weak ties" or acquaintances--those with whom the individual has infrequent contact (7% of our sample in nuclear waste, 10.5% in photovoltaics). A third circle, larger still, consists of those known by "reputation only," perhaps through reading the work of a researcher or through friends--our category of "heard of but no contact" (10.5% in nuclear waste, 14.1% in photovoltaics). Beyond this is a large number of system participants unknown to the respondent, individuals who are potential contacts and may be brought into his/her personal network in the future. For the relatively elite group represented here, this "unknown periphery" is on the average over three-fourths of the system in radioactive waste and over two-thirds of the system in photovoltaics.

Formal Communication Events

Indicators of formal modes of communication are presented in Exhibit 4.2. In a large-scale enterprise, "formal" may have two meanings. First, there is the system of technical reports, memoranda, and plans--all variants of the written word. Second, there is the system of patterned interpersonal interaction estalished by lead agencies and professional organizations. The latter is our concern here.

Some indication of the effectiveness of formal communication structures is evident in

Exhibit 4.2 Formal System Events and External Participation

	Nuclear Waste	Solar Cells	Full Sample
Percent participation in a project integration meeting, workshop, or periodic review meeting for Department of Energy contractors during the past three years	80%	86%	83%
Member of a government or professional committee or advisory group concerned with the field during the past three years	50%	40%	45%
Average number of meetings (government sponsored or professional society) during the past three years			
All meetings	16	16	26
Waste/Cells only	9	11	10
Percent who had frequent dealings during the past year (8-10 times or more) by phone, letter, or in person with: (excluding same organization)			
The media	13%	9%	11%
Nonscientists	16%	23%	19%
Percent serving as a paid or unpaid consultant within their professional field during the past three years	43%	52%	47%
Percent giving expert advice to a public-interest group within the past three years	47%	43%	45%

responses to questions on meetings and committees. Eighty-three percent of the total sample had participated in a project integration meeting, workshop, or periodic review meeting for Department of Energy contractors in the last three years. Nearly half had served on a government or professional advisory group.[6] Finally, respondents were asked how many professional meetings within the field they had attended within the past three years. Exhibit 4.2 shows an average of nine such meetings were attended by respondents in nuclear waste, about eleven for those in photovoltaics. This averages to about one meeting, sponsored by government or professional societies, every three or four months.

Writers, journalists, and television reporters have consistently been interested in the progress of technology in nuclear waste and photovoltaics. About one-tenth of the sample reported frequent contact with the media. One implication of the technical system concept is the importance of relationships with nontechnical personnel (Shrum, 1984). Exhibit 4.2 shows that about one-fifth of the respondents had frequent dealings of a professional nature with nonscientists outside their own organization, providing some descriptive support for this claim.

Communication also involves consulting and advisory activities. Photovoltaic actors were more likely to be involved in consulting. Fifty-two percent had served as paid or unpaid consultants during the past three years, as contrasted with 43% in nuclear waste, a difference of nine percentage points. Nuclear waste actors were slightly more likely to be involved with public-interest groups. Nearly half (47%) had given expert advice to a public-interest group within the past three years.

To this point evidence for a variety of communicative behaviors with different sectors and groups, both inside and outside the organization, has been presented. Whether the rates are evaluated as high or not depends on the standard one chooses to use. The contrast between nuclear waste and photovoltaics did not reveal large differences on most indicators. One recent study of agricultural scientists found that communication outside the organization took place monthly on the average, judging this rate relatively low (Busch and Lacy, 1983). However, if the discussion is "useful to work in the field," as in the item used here, this rate may be fairly significant.

Formal and Informal Linkages

The second question pertains to the relationship between formal system events and informal communication. Exhibit 4.2 shows that a significant proportion of the sample had participated in DOE meetings and governmental or professional advisory groups. In Chapter Three, the use of meetings to stimulate interaction among researchers in the photovoltaic private sector was discussed. However, formally structured interactions are generally thought to be important for innovation in any technical system. One function of these meetings is to facilitate communication directly through creating situations in which researchers and program managers present plans or results of work and discuss problems and approaches. A second function is to promote communication indirectly through the creation of social relationships. Once a tie is established researchers with common interests may come to exchange information and assistance with one another on an informal basis as needs and opportunities arise.

Linkages formed at institutionalized system events persist over time and serve as important channels of information for technological development.[7]

Both functions were recognized by informants:

> Almost every year [our program has an] information meeting with DOE, NRC, subcontractors, and other experts...[who] aren't funded. These meetings are an important means of communication....Communication is good overall but not as good as it could be. The sheer size of the field is a problem.... Conferences in general are quite important in communication....The most valuable aspect of meetings is the informal contacts with people.--national lab researcher in nuclear waste

> IEEE [the largest conference in photovoltaics] is by no means the best technical meeting.... There are contract review meetings held once a year at SERI. They are subject-focused, vary in size according to how many work in the area, and function as a general review. [There's] not too much technical information presented, mostly "awareness" for latter exchange.--lead center researcher in photovoltaics

Although opinions vary regarding the quality and timeliness of information presented at these formal events, there seems to be a consensus that the establishment or maintenance of personal interactions is an important reason for their frequency. As one NASA manager put it, "they're carefully structured--lots of thought goes into getting people together in the halls."

We are interested here in whether there is a tendency for those who are involved in formal system events to exhibit stronger informal communication patterns as well. Is there, in empirical terms, an association between indicators of formal system participation and indicators of discussions or professional dealings with other actors in the system? If not, we should be sceptical of this function of meetings.

Exhibits 4.3 and 4.4 show the results of cross-tabulating measures of formal and informal communication for nuclear waste and photovoltaics respectively. In this analysis only researchers are included.[8] Percentages in the table represent the proportion of those who attended Department of Energy meetings or were members of advisory committees who scored high on each indicator of informal communication.

The first two rows of each table show the results of "other-reported" network measures, the most reliable and valid indicators of communication. Clearly, those who have attended project integration meetings, workshops, or periodic review meetings have contact with a much larger proportion of both the entire sample and the (subset of) researchers than those who have not. The results from both systems are similar, with differences of thirty to forty-five percentage points. Likewise, those who have served on government or professional advisory committees are more likely to have larger personal networks, with differences in the same range. All eight relationships are statistically significant.

The remaining measures are similar to those in Exhibit 4.1. The results are not quite as striking, but all relationships except one are in the expected direction (discussions with persons at other organizations in photovoltaics) and most are statistically significant.[9] High rates of contact with specific sectors (in particular government agencies) are associated with formal participa-

Exhibit 4.3 Relationship between Formal Events and Informal Communication--Nuclear Waste*

Indicators of Informal Communication:	Percentage with High Levels among Attenders**	Percentage with High Levels among Nonattenders	Percentage with High Levels among Members***	Percentage with High Levels among Nonmembers
(1) Contact with all respondents in sample	54.9%	10.5% (p. 001)	70.2	23.6 ($p<.001$)
(2) Contact with researchers in sample	53.8	10.5 ($p=.001$)	66.7	25.5 ($p<.001$)
(3) Useful discussions with persons in waste at same organization	88.9	68.4 ($p=.04$)	96.4	72.7 ($p=.001$)
(4) Useful discussions with persons in waste at other organizations	81.1	52.6 ($p=.01$)	83.9	69.1 ($p=.065$)
(5) Useful discussions with persons outside field at other organizations	38.9	36.8 ($p=.87$)	42.9	34.5 ($p=.369$)
(6) Contact with government agencies	81.1	47.4 ($p=.004$)	89.3	60 ($p<.001$)
(7) Contact with business and industry	73.3	42.1 ($p=.008$)	73.2	61.8 ($p=.20$)
(8) Contact with universities	58.9	47.4 ($p=.357$)	73.2	40 ($p<.001$)
(9) Contact with national laboratories	87.8	57.9 ($p=.004$)	89.3	74.5 ($p=.044$)

*Only researchers are included (national laboratories, universities, and private firms). N=109-112. Items (1) and (2) have been dichotomized at the median value of the distribution. Both frequent and infrequent contact have been included. See Exhibit 4.1, items 1-3 and 7-10 for question wording.

**Exact wording of the question was "During the past 3 years have you participated in a project integration meeting, workshop, or periodic review meeting for Department of Energy contractors?"

***"During the past 3 years have you been a member of a government or professional committee or advisory group concerned with photovoltaics?"

Exhibit 4.4 Relationship between Formal Events and Informal Communication--Photovoltaics*

Indicators of Informal Communication:	Percentage with High Levels among Attenders**	Percentage with High Levels among Nonattenders	Percentage with High Levels among Members***	Percentage with High Levels among Nonmembers
(1) Contact with all respondents in sample	54.5%	20% (p=.005)	67.4	38.6 (p=.002)
(2) Contact with researchers in sample	54.5	25 (p=.015)	71.7	37.3 (p<.001)
(3) Useful discussions with persons in field at same organization	90.8	78.9 (p=.215)	95.6	85.4 (p=.112)
(4) Useful discussions with persons in field at other organizations	72.7	73.7 (p=.93)	78.3	69.5 (p=.286)
(5) Useful discussions with persons outside field at other organizations	40.7	36.8 (p=.749)	54.3	32.5 (p=.016)
(6) Contact with government agencies	92.7	60 (p<.001)	95.7	83.1 (p=.039)
(7) Contact with business and industry	85.5	75 (p=.344)	100.0	74.7 (p<.001)
(8) Contact with universities	76.4	55 (p=.048)	80.4	68.7 (p=.15)
(9) Contact with national laboratories	70.9	35 (p=.002)	78.3	57.8 (p=.02)

*Only researchers are included (national laboratories, universities, and private firms). N=126-130. Items (1) and (2) have been dichotomized at the median value of the distribution. Both frequent and infrequent contact have been included. See Exhibit 4.1, items 1-3 and 7-10 for question wording.

**Exact wording of the question was "During the past 3 years have you participated in a project integration meeting, workshop, or periodic review meeting for Department of Energy contractors?"

***"During the past 3 years have you been a member of a government or professional committee or advisory group concerned with photovoltaics?"

tion in both systems (rows 6-9). Formal participation is associated with useful discussions within one's organization in nuclear waste but not in photovoltaics (row 3). One explanation for this may be the larger size of organizations in nuclear waste. When the number of researchers in a field within a national laboratory becomes large, individuals in other sections may become functionally distant and require the same sort of mechanisms to increase informal contact as between researchers in other organizations. In photovoltaics, smaller organizational populations do not require such mechanisms so there is no relationship between formal participation and frequency of informal discussions within one's own organization.

The fifth row in each table shows a lack of significance on three out of four associations between attendance at formal events and discussions with persons outside the field. This supports the general notion of the function of formal events, because they would not be expected to include nonsystem members. Overall rates of communication are not increased by formal participation, only communication within the system itself.

The major anomaly in the table is the absence of an association of formal participation and useful discussions with persons at other organizations in photovoltaics. Perhaps this is due to deficiencies of the measure itself or perhaps to the large number of private-sector actors in the photovoltaic system. These researchers would not be expected to communicate *as* freely as those in other sectors, a matter to which we return in the final section of the chapter.

Intersectoral Communication

The association between participation in formal system events and high levels of informal communication established for researchers augments the description of the formal organization of these fields by suggesting the influence of the administrative component. However, more direct evidence of the role of administration would be desirable. One means of assessing this role is to examine the relation between the administrative sector and each other sector individually. This may be accomplished through an analysis of the sectoral location of each of the reported linkages between actors in the sample.

The distribution of social associations across sectors is important for other reasons as well. One of the principal tasks of structural sociology is the determination of the extent to which social associations span group boundaries (Blau, 1977). Further, there has been considerable speculation about the appropriate relations between university and industry, industry and government, and university and government (Halloman, 1979; Praeger and Omenn, 1980). This has been accompanied by a number of case studies which are primarily concerned with institutional relations and policy recommendations. The consequence has been an almost complete neglect of the social associations which constitute the bedrock of institutional relations.[10] In this section descriptive models are presented which use rates of interpersonal contact as an indicator of the strength of relationship between sectors.

Once again, network measures are used to derive information on the relations between sectors. Sectoral location is simply defined as the type of organization to which each respondent belongs: government agency, national laboratory, private firm, university, public-interest group, and--in the nuclear waste system only--"policy" organizations.[11] To the extent that these organizational types are different--as indeed a large body of organizational analysis suggests they are--sectors constitute interactional boundaries. That is, interaction across sectoral boundaries is crucially important for the operation of the technical system with respect to the flow of information, favors, and legitimation. Government administrators and program managers are crucial providers of resources and information regarding program needs and directions for individuals in private firms and national laboratories. For private researchers, universities are sources of technical information not governed by proprietary restrictions. Private firms and national laboratories offer subcontracts and facilities to researchers. Virtually all such exchanges involve interpersonal linkages between representatives of these kinds of organizations.

Construction of Density Models

Sectoral density models, as in Exhibits 4.5 and 4.6, are a means of indicating relative levels of interaction among groups. Three steps are involved in their construction. First, a binary adjacency matrix is created from the item on communication introduced earlier in the chapter. The matrix is square, eliminating nonrespondents. Each row represents the reported contacts of one respondent. The corresponding column represents the reported contacts of others with that respondent. In each cell of the matrix a value of one is entered if frequent or infrequent contact is reported by actor A to actor B. A value of zero is entered if no contact is reported.

Exhibit 4.5 Communication between Sectors--Nuclear Waste System*

	Gov.	Labs	Priv.	Univ.	Pub.	Pol.	Ties** Given
Government (n=10)	.583	.229	.077	.195	.150	.179	.191
National Labs (n=56)	.248	.187	.040	.104	.071	.089	.126
Private (n=31)	.116	.064	.026	.035	.022	.032	.049
University (n=22)	.127	.078	.012	.101	.038	.057	.066
Public-Interest Groups (n=6)	.067	.065	.054	.121	.538	.174	.103
Policy (n=24)	.212	.103	.039	.125	.208	.197	.118
							Overall Density
Ties Received	.199	.128	.036	.099	.097	.097	.103

*Cell values represent proportion of possible interpersonal relations actually reported. Both frequent and infrequent ties are included. Relationships between members of the same organization have been *excluded*.

**Row marginals indicate the proportion of possible contacts *reported* by a given sector to all sectors. Column marginals indicate the proportion of possible choices *received* by a given sector from all sectors.

Exhibit 4.6 Communication between Sectors--Photovoltaic System*

	Gov.	Labs	Priv.	Univ.	Pub.	Ties** Given
Government (n=9)	.679	.250	.257	.380	.044	.285
National Labs (n=36)	.281	.114	.117	.155	.028	.132
Private (n=63)	.326	.134	.133	.161	.006	.147
University (n=31)	.376	.118	.127	.190	.019	.149
Public-Interest Groups (n=5)	.133	.033	.022	.058	.444	.050
						Overall Density
Ties Received	.327	.130	.132	.175	.028	.149

*Cell values represent proportion of possible interpersonal relations actually reported. Both frequent and infrequent ties are included. Relationships between members of the same organization have been *excluded*.

**Row marginals indicate the proportion of possible contacts *reported* by a given sector to all sectors. Column marginals indicate the proportion of possible choices *received* by a given sector from all sectors.

This matrix is not symmetric, since respondents may not be consistent in reporting contacts. Such information is itself useful as an indicator of relative status. Second, the rows and columns of the matrix are shifted so that individuals in each sector are grouped together: all respondents from government program offices are grouped, followed by those from national laboratories, and so forth. This creates a series of submatrices within the large matrix, each of which shows all of the reported linkages (or nonlinkages) between members of a given sector with members of another sector (or members of a sector with each other, in the case of submatrices on the diagonal).

The final step in creating the sectoral density model is the calculation of submatrix density values. A submatrix "density" is that proportion of *possible* linkages (the count of all cells in the submatrix) which *actually occur* (the count of ones in the submatrix). Therefore, the cell values in Exhibits 4.5 and 4.6 may be read as the relative frequency of interaction or communication between and within sectors, ranging from a low of zero (no interaction between any members of two sectors) to a high of one (all individuals are linked directly). Essentially, the use of submatrix densities is a way of reducing the information in a large matrix to manageable form. What began as a matrix of 150 rows and columns is reduced to a matrix of five rows and columns.

Hypotheses on Contact and Status

We begin by stating hypotheses based on the theoretical and historical materials presented in Part One and the results of the bibliographic search concerning the sectoral distribution of output for each system. The hypotheses are stated in terms of the density values representing rates of association between sectors in these models. Of course, there is often much to be found in examining models that we are not looking for. But it is wise to have some notion in advance of what is likely to be there.

> (1) Intersectoral relations involving government will display higher densities than those not involving government. In the nuclear system, overall rates of contact should rank government, national laboratories, universities, and private firms in that order; for photovoltaics, overall rates of contact should rank government, private firms, universities, and national laboratories in that order.

This hypothesis embodies two predictions based on the functional roles of the sectors within each system. Government agencies perform control and coordination functions and thus should give and receive more linkages than other sectors in each system. Higher density values are therefore expected in cells involving government.

Overall rates of contact are indicated by the values in the column marginals for each system. These totals express the proportion of all possible choices actually *received* by a given sector from all sectors.[12] The greater the value of the column marginal, the higher the overall rate of communication for that sector within the system as a whole.

The hypothesis predicts that the primary research sector should rank second behind government in each system (private firms in photovoltaics, national laboratories in nuclear waste). Given their dominant role in the production of technical output, it is probable that they give and receive a large share of communication linkages. Universities play an important role in the nuclear system, but are much more active in photovoltaics, given the more basic nature of photovoltaic research. As a source of "generalized" knowledge academic researchers are expected to be a source of information and legitimation for both systems. Finally, laboratories in photovoltaics and private firms in nuclear waste are "peripheral" sectors in that they produce a small proportion of the output and are not expected to be as well integrated into the communication network as the other research sectors.

> (2) The structure of the nuclear waste system should reveal a "public core" with high densities between and within laboratory and governmental sectors and relatively low densities for other sectors. The distribution of linkages in photovoltaics will be relatively more uniform throughout the research sector.

The public character of radioactive waste and the concentration of output in this system lead to the expectation that the national laboratories and governmental sector will form a tightly integrated "public core" and a periphery consisting of academic and private sectors. Most of the social relationships will tend to fall within this core--thus, the four submatrix densities should be relatively higher than those for the other research sectors. In photovoltaics, while there may be a core and periphery (in terms of active and less active researchers), this structure does not seem to follow sectoral lines. The distribution of ties should be weighted toward the governmental sector, but within the research sectors it is likely to be uniform.

(3) The ratio of linkages received to linkages given should be highest for the academic sector and lowest for the public-interest sector.

This is a hypothesis on the relative status of sectors within the systems. Sociometric and small-group studies have shown that higher-status actors tend to be the recipients of large numbers of choices, relative to the number of choices made. Conversely, low-status actors tend to choose others more than they are chosen. Given this background, the ratios of column marginals to row marginals in an adjacency matrix may be used as an indicator of status (Beniger, 1980). Traditionally, universities are contexts with higher status than other technical organizations due to the higher prestige of basic research and the greater autonomy provided by academic research careers. Additionally, academic researchers are often drawn in on review panels and planning boards as independent sources of expertise. Solicitation of advice and general sectoral status lead to the expectation that universities will emerge as the highest status sector within both technical systems.

At the other extreme, public-interest groups are not central to the technical tasks of the system and although they play an important role in the political direction of the system, they are not highly regarded by most researchers and governmental actors, as we saw in Chapter Three. Often public-interest advocates are considered an irritant or drag on the development of technical solutions, imposing impossible levels of certainty on health and environmental criteria and making annoying or unreasonable requests for information. They should display relatively low ratios of ties received to ties given.

Exhibits 4.5 and 4.6 present sectoral density models for nuclear waste and photovoltaics which tend to support these hypotheses. The overall density of ties is higher in photovoltaics than in nuclear waste, as indicated by the table averages. Fifteen percent of possible relationships among actors in the photovoltaic system and 10% of those in the nuclear waste system were actually reported. Strong support is evident for hypothesis (1) regarding the role of government as an integrating sector. In photovoltaics the first row and column of Exhibit 4.6 show that ties between government and all research sectors are stronger than ties between or within any research sector. In nuclear waste, this is true except for the low government/private-sector density as reported by government and the high laboratory densities.

These strong within-sector relations among the national laboratories in nuclear waste support the second hypothesis which specifies the existence of a "public core" consisting of government and laboratories and excluding universities and private firms. Almost 19% of possible linkages among laboratory individuals were reported by respondents.[13] In contrast, ties between private firms, universities, and laboratories range from 4% to 10% of those possible. This high-density core together with relatively low densities throughout the remainder of the research sectors is absent in photovoltaics. Within the research sector ties are distributed uniformly (Exhibit 4.6) as predicted, ranging from .114 to .19. Considering ties received by laboratories and private firms only, the range is only .114 to .134. The highest level of contact within the research sectors in photovoltaics is among university researchers, where rates of 19% are roughly equivalent to those within the national laboratories in nuclear waste.

Hypothesis (1) also provided an expected rank ordering of marginal densities indicating the relative number of linkages received for each sector. For nuclear waste this ordering was (from highest to lowest) government, laboratories, universities, and private firms. This prediction is confirmed, taking column marginals, with values of .199 for government, .128 for laboratories, .099 for universities, and .036 for private firms.[14] For photovoltaics, the predicted ordering of sectors was government, private firms, universities, and laboratories. However, the data show a significant reversal for industrial and university researchers. Column marginals in Exhibit 4.6 show government received the highest proportion of selections (.327), followed by universities (.175), private firms (.132), and laboratories (.130).

The final hypothesis specified expected ratios of column to row marginals for academic and public-interest sectors as an indication of the relative status of various sectors composing the system. In nuclear waste the ratio for universities is 1.5 (.099/.066), which exceeds the ratio for government (1.042), laboratories (1.016), and private firms (.735). In photovoltaics the university ratio is again the largest (1.174), compared with 1.157 for government, .985 for laboratories, and .898 for private firms. The academic sector appears to have the highest status, at least in the sense of receiving more linkages than it sends.[15] With respect to public-interest groups, the photovoltaic system shows them to have the lowest ratio of ties received to ties given (.56). However, in the nuclear system this is not the case. Here, private firms (.735) and other policy actors (.822) score lower than public-interest advocates (.942).

Overall the most striking finding from Exhibits 4.5 and 4.6 is the clear indication that the social structure of the two systems is similar with respect to the role of government as the integrating sector and different with respect to interaction within the research sector. Though a small sector in absolute terms, program managers in federal agencies and lead centers maintain extensive associations with all system components, providing "bridges" or indirect ties between actors where those actors are not themselves directly linked. Each sector in both systems reports more associations with government than with any sector with the exception of public-interest groups in nuclear waste. This sector reports a greater proportion of linkages to both academic and policy sectors than to government.

The greater importance of the public-interest sector to the nuclear waste system is shown by the difference in the proportion of choices received by that sector in the two systems. Not only is the absolute proportion greater in the nuclear waste system (.097 to .028), but more importantly, the proportion is very close to the table average, signifying that the sector is relatively integrated into the system. In photovoltaics the public-interest sector is selected less than 3% of the time, whereas the table average is nearly 15%.

Another difference lies in the relative status of the public-interest sector, indicated by the ratio of ties received to ties given. In photovoltaics this ratio was the lowest of any sector in the system. However, in nuclear waste it is ranked more highly than private or policy sectors. An indication of the reason for this lies in the examination of specific inter-sectoral ratios, particularly with respect to the government/public-interest group relationship. Governmental actors report 15% of possible contacts with public-interest actors, while public-interest actors report only 6.7% of possible ties with government. The most plausible explanation of this is the greater need for legitimacy in the nuclear waste system, sensitizing managers to the demands and importance of public-interest groups.

Government program managers are often an embattled group with low credibility among the general public for reasons outlined in Part One. Linkages with public-interest groups serve to symbolize their commitment to open dissemination of information and receptivity to alternative viewpoints and criticism. Though these relations may often be viewed as a burden in terms of time and energy, they are valued and promoted as a means of countering the long-term criticisms of secrecy and elitism directed at federal nuclear programs. The outcome is a subtle but distinct dependency of government on public-interest groups in the nuclear system, leading to their relatively higher status.

A significant structural difference between the two systems also exists with respect to the extent of social interaction within the research sector. In nuclear waste, laboratories are more highly integrated than other sectors and intersectoral linkages are not large. University/laboratory ties are the only ties which approach the table mean. In photovoltaics both intra- and intersectoral linkages are strong. In both systems, academic researchers seem to be much more significant in the communication network than would be predicted by their productivity relative to other sectors (Exhibit 3.5). As hypothesized there is a sectorally defined core within the nuclear waste system, while communication is relatively more uniform within photovoltaics. This may indicate that the governmental function of coordination is more significant within the former system, given the absence of direct linkages.

Rates of Exchange

We have addressed the intersectoral relationships in the two systems by means of models showing the extent of interpersonal linkages between and within sectors. It was assumed that frequent or infrequent professional dealings implied communication, and that this communication was important to the innovation process within the system because of its potential for information and resource flows. Of course, it is to be expected that communication ties have varying degrees of efficacy.

Contrary to expectations based on the notion of a "primary research sector," universities were observed to have a more important role in photovoltaics than private firms as indicated by the proportion of linkages received (Exhibit 4.6). One plausible explanation for this reversal is interorganizational competition within the private sector in photovoltaics. As described in Chapter Three, private firms must be motivated to participate in a system which requires disclosure of research results. From the standpoint of the firm, it is advantageous to use federal funds for routine, nonsensitive research, and internal funds for sensitive research (Eckhart, 1978).

However, given proprietary restrictions on the dissemination of research results, an organizational requirement imposed on the verbal behavior of privately employed scientists, less information is available to the *system* relative to the resources invested than in other sectors. University scientists, on the other hand, are given virtually complete freedom to associate and discuss results with whom they choose.[16]

While university scientists do have higher rates of interaction than private scientists in photovoltaics, the rate of internal, private-sector communication is by no means insignificant. Confirmation of the concept of restricted information flow for the private sector requires a more direct measure of exchange or professional assistance.

The questionnaire item on professional dealings selected that subset of the sample with which each respondent had frequent or infrequent contact. Following this he/she was asked: "For each of the individuals with whom you have had dealings during the past three years please indicate whether you have assisted that person." A list of various forms of professional assistance (giving information, supplying a sample, etc.) was provided for those who requested definitions of "assistance." This measure is an indicator of an *exchange* relationship rather than simple contact or association. Such a relationship is more intense, implying obligations, commitment, and a more lasting tie. Like contact, it may be used in models of intersectoral relations. However, instead of taking the proportion of possible contacts which occur, we now take the *proportion of actual contacts which involve exchange*. What was previously the numerator in calculating sectoral contact density is now the denominator in calculating sectoral exchange models.

Using this measure we can compare the relative rates of exchange for the private sector in photovoltaics with other sectors in that system. If the behavior of industrial scientists and engineers is affected by these organizational restrictions, we would expect these to be reflected in lower rates of exchange. The first hypothesis on exchange rates is specific to the photovoltaic system:

> (4) In systems oriented to the production of private goods, the ratio of assistance given to total contacts will be relatively lower for the private sector than for other sectors and lowest for relationships within the private sector.

The second hypothesis applies to both systems and follows from the conception of a technical system as a network of actors with functionally differentiated components.

> (5) The ratio of assistance given to total contacts will be relatively higher for administrative sectors than for nonadministrative sectors.

Given the resources available for distribution by the administrative component and its integrative function for the system as a whole we would expect members to perform relatively more favors. Such favors are more likely to be responses to requests for information on program needs and requirements than direct grants of funds. The central position of program administrators in the network of actors gives them a high level of awareness about system potentials and events, including the existence of research funds, which researchers are investigating similar problems, where needed samples can be acquired or analyses performed, and who is likely to be interested in purchasing certain skills and capabilities. In both the nuclear waste and solar cell systems, national laboratories and governmental agencies perform administrative functions due to the program decentralization described in Chapter Three. It is therefore expected that both sectors will have relatively high ratios of assistance given to total contact, with government displaying the highest ratios.

The cell values in Exhibits 4.7 and 4.8 present proportions representing the number of relationships involving reported assistance divided by the number of relationships involving frequent and infrequent professional dealings. Exhibit 4.8 supports the hypothesis of lower private-sector exchange in the photovoltaic system. Forty-four point seven percent of actual relationships among researchers in the private sector were reported to involve professional assistance, lower than any other cell in the matrix. In fact the three lowest values in the table were reported by the private sector in its relationships with government, universities, and internally. High rates of assistance are reported by laboratory and governmental personnel--six of the eight cells in the top two rows are above the table median. University researchers also report high rates of assistance to government. The nuclear waste system (Exhibit 4.7) shows a great deal of similarity, although exchange

Exhibit 4.7 Exchange between Sectors--Nuclear Waste System*

	Gov.	Labs	Priv.	Univ.	Assistance Given
Government (n=10)	.643	.562	.667	.372	.547
National Labs (n=56)	.612	.563	.514	.477	.554
Private (n=31)	.444	.450	.435	.375	.438
University (n=22)	.643	.469	.125	.455	.477
					Overall Exchange
Assistance Received	.593	.538	.504	.444	.529

*Cell values represent the proportion of social contacts involving *assistance* or exchange. Relationships between members of same organization have been excluded.

internal to the private sector is not the lowest value in the table. Since it is a noncompetitive system, private researchers are not as confined by organizational restrictions. Relatively high values are evident within the public core of government and laboratory researchers. As in Exhibit 4.8, university researchers report high levels of assistance to government. Although the highest and lowest values in the table are assistance to the private sector by government and universities respectively, these figures are based on extremely few relationships due to the low rates of contact involving the private sector as reported in Exhibit 4.5.[17] Low rates of assistance are reported by the private sector (all four values below the median), while

Exhibit 4.8 Exchange between Sectors--Photovoltaic System*

	Gov.	Labs	Priv.	Univ.	Assistance Given
Government (n=9)	.789	.580	.568	.557	.580
National Labs (n=36)	.648	.652	.620	.578	.620
Private (n=63)	.503	.576	.447	.487	.514
University (n=31)	.648	.538	.536	.512	.549
					Overall Exchange
Assistance Received	.587	.584	.545	.523	.554

*Cell values represent the proportion of social contacts involving *assistance* or exchange. Relationships between members of same organization have been excluded.

assistance *to* universities is relatively slight as well.

The models in Exhibits 4.7 and 4.8 offer substantial support for hypotheses (4) and (5). First, the governmental and laboratory sectors tend to report the highest levels of assistance given relative to their social associations. Second, the private sector is involved in fewer grants of assistance than other sectors in both systems, and is particularly low for relationships within the private sector in photovoltaics, as expected.

It is important to stress that alternative interpretations of these results are possible. Although reports by system personnel are likely to be the best source of information regarding exchanges within these fields, the

possibility remains that responses are affected by different norms concerning *reports* of exchange. That is, perhaps governmental and laboratory personnel report high levels of assistance because of the norm of service applicable to the public sector generally rather than because of their greater opportunities to lend such assistance. What argues against this interpretation are the low rates of assistance by academic researchers, whose norms of free and open communication would seem to favor such responses as well.

Summary

The analysis in this chapter has enriched our understanding of collective and private systems by adding a quantitative dimension to the historical and organizational views of nuclear waste and photovoltaics as large-scale technological enterprises. Of fundamental importance is the possibility of using a combination of standard survey methodology and modern network techniques to provide information on the social structure of technical systems.

Substantively, the results of both standard survey and network measures suggest two relatively integrated social systems with significant levels of cross-sectoral communication. A high proportion of researchers participate in formal system events. This participation is associated with high levels of informal communication.

The problem of intersectoral relations was given an empirical interpretation in terms of rates of social associations within and between sectors. Governmental actors tended to have the highest rates of contact with other sectors for both systems. However, the nuclear waste system was structured around a public core consisting of governmental and national laboratory actors, with relatively low rates of linkages between other research sectors. In contrast, relationships within the photovoltaic system were not mediated by government, and linkages within the research sectors were fairly uniform. The public-interest sector exhibited stronger ties with the radioactive waste system, primarily through governmental and university linkages. Finally, an examination of exchange relationships within the administrative and research components revealed high rates of assistance given by government and laboratory sectors and low rates by the private sector. Intra-sectoral competition within photovoltaics was implied by low rates of exchange within the private sector.

Chapter 5

Research Performance in Technical Systems

The final chapter of the argument relates the processes and structures which have been described at the broader system and sectoral levels to the performance of the individual researcher. It is an attempt to provide another piece of the answer to the question raised at the outset of Chapter Four: what difference does the nature of the technology under development make to the process of innovation?

Thus far we have seen that the notion of a collective good provides some understanding of the observed differences in output, sectoral involvement, administrative, and communication structures of the nuclear waste and photovoltaic systems. Historical factors were crucial to the establishment of an organizational locus for photovoltaics as national energy policy and the shifting costs of fuels led to a terrestrial photovoltaic program and increasing private investment. The radioactive waste issue occupied a critical position for the nuclear energy industry and its federal counterpart. As the perceived risks and visibility of the problem increased, the research system grew in size and scope. Currently, the national laboratories are the primary research sector in nuclear waste, depending exclusively on federal funds to carry out the research program. Thus, the use of these technologies for collective or private benefits conditions distinctive interorganizational structures in each field.

The problem addressed in this chapter is a straightforward extension of hints derived from the preceding chapters. In Chapter Two, a stronger association between the innovativeness of a subfield and its level of activity seemed to imply more effective control by the administrative sector in nuclear waste. In Chapter Three, analysis of the output of the nuclear waste system established the dominance of national laboratories. In Chapter Four the communication structure in nuclear waste was clearly organized around the national laboratories and governmental sectors. It is thus plausible that the determinants of performance for the individual researcher depend on the type of system. In systems involving public goods, given the relative monopoly of funds and expertise by the government, the administrative component of the system is likely to possess a larger degree of influence than in systems where private goods are at stake.

A large number of studies of scientific and technological performance have shown that the position of an individual within a network of relations is correlated with the observed level of productivity, measured in various ways (Pelz and Andrews, 1976; Crawford, 1971; Breiger, 1976). In general, those individuals who are more central to the communication network--giving and receiving large numbers of sociometric choices (as described in the previous chapter)--are most likely to exhibit high levels of technical performance.[1] This finding is based on studies of firm-based innovation and basic scientific specialties. It has yet to be tested with large-scale enterprises such as nuclear waste and photovoltaics. Here it will be used as a standard with which to compare determinants of performance in technical systems.

Of course, there are other reasons to be interested in performance. For the research manager, whose goal is the maximization of performance by individuals and research teams, it is important to know the factors which are conducive to high-quality technical work. Size and quality of the organizational environment, among other variables, may influence the level of performance irrespective of network position.

After a brief discussion of the measurement of performance in technical systems, evidence is presented on the relationships between network position, organizational context, and performance. Simple models are developed to show that nuclear waste and photovoltaics differ in terms of the type of social relationships which are beneficial for high levels of technical work. Next, it is shown that these models do not work equally well for researchers in all sectors. Finally, performance in the primary research sectors for each system is examined.

Measures of Performance in Technical Systems

The measurement of performance is surely one of the most difficult tasks in the sociology of science and technology. It is possible, on the one hand, to count publications (as in the sampling phase of the present study) to indicate the productivity of an individual researcher or organization. In basic science, there are good arguments for the use of citation counts as an indicator of the extent to which work has been used or acknowledged by others. However, the best measure of a scientist's work is really quite clear, difficult though it may be to obtain in practice. It is the evaluation of that work by knowledgeable peers--those who are competent and familiar with it. There may be those who will object that the quality of a contribution has some independent or "objective" sense apart from assessments made by other scientists. But it is difficult to imagine a plausible criterion other than the judgment of competent peers, unless it is the derived knowledge of the historian of science, making the same assessment in the light of time. There is no getting around the fact that the touchstone for technical work is the opinion of other members of the technical community.

Two principal measures of research performance were used in this study. Each was constructed using the responses of a subset of the sample to assess the contributions of a given researcher. The first was based on an open-ended item asking: "In your opinion, which persons have contributed the most significant technical innovations in the field of (radioactive waste/photovoltaics) during the *past three years?*" Spaces were provided for up to seven individuals and organizations. These responses were used to distinguish a relatively elite group of actors according to the absolute number of nominations received.

A second, fixed-format item was appended to the roster portion of the questionnaire described above. Respondents were asked to review the roster of individuals and "*Name* only those individuals with whose *work you are most familiar*." Having narrowed the sample to this subset, the interviewer then went on, "Now I would like you to indicate your assessment of the contribution to important new developments in the field within the past three years for each of the individuals you have just mentioned." The evaluation was made on a six-point scale, with high values indicating large contributions to the field. Scores of all persons who rated a given individual were averaged as a measure of performance from the perspective of knowledgeable persons within the technical system.[2] Respondents rated an average of 11.6 individuals in radioactive waste and 19.2 individuals in photovoltaics.

It is noteworthy that these two measures of performance do not directly reflect the counted productivity of researchers, as indicated by the bibliographic search. The correlation coefficient for the relationship between rated performance and the count of publications is .37 in photovoltaics and .21 in radioactive waste.[3] For nominated performance the correlation with publications is .37 in photovoltaics and .31 in radioactive waste. We may therefore infer that while there is a slight tendency for those who have produced many articles and reports in these systems to produce technical contributions of higher quality, it is by no means the case that knowing a researcher's level of productivity is a substitute for knowing how his work is evaluated. That is, there are substantial numbers of researchers who produce fewer publications of good quality and many who produce large numbers of publications whose rated contribution is not great. In order to assess individual research contributions, direct evaluations must be made: counting publications is no substitute.

More puzzling is the relatively low correlation between nominated and rated performance: .30 in nuclear waste and .41 in photovoltaics. If both of these measures are indicators of the same underlying concept (evaluations by knowledgeable peers of contributions to technical knowledge), we would expect them to be more strongly associated.[4] Perhaps the difference in measurement techniques corresponds to distinct concepts of performance. Though both reflect collegial evaluations, in one case the assessment is made of one's work *relative to a set of other individuals*; in the other, a single probe elicits names corresponding to the social role of an "innovator." Nominated performance may not indicate the deliberate evaluation of a researcher's work so much as a social definition or "label" representing a researcher's reputation in the field. Indeed, most of us could name a number of individuals in our own fields whom we accept as "eminents" based on colleagues' judgments, but whose work we do not know firsthand.

To test this notion of an innovative label it is important to know whether those that are being nominated as innovators are individuals whose work is *known by* the nominator. If nominations are made at least partly based on social reputational factors, we would expect to find that individuals whose work is not well known may be nominated as well. If, however, nominations represent the *most* innovative researchers from among those with whose work the respondent is familiar, then nothing beyond the respondent's "cutoff" level for innovators--where he draws the line for the most elite researchers--can be inferred, a thoroughly unimportant fact. In operational terms, we need to know whether respondents nominate those whom they also feel sufficiently well qualified to rate.

Taking only those nominations of individuals who also appeared in the sample, the proportion of those who were nominated but *not* rated was calculated. In radioactive waste 30% (22 of 73) and in photovoltaics 29% (76 of 260) of the nominations were not among those with whose work the respondent was most familiar. Thus, nearly a third of nominations are to researchers not well known to the nominator. Although it is not yet clear what specific dimensions are important to the labelling process, it does seem that extra-technical factors enter into attributions of innovation. By analyzing nominated performance separately as an indicator of the extent to which a researcher has been defined as an innovator, one might begin to determine what these factors are.

Effects of Contacts and Context on Performance

Theories of Big Science or "industrialized research" have made much of the increasing technical requirements of research (Price, 1963; Ravetz, 1971). "Environmental" or "resource" factors are important facilitators of technical performance, which is one reason for the emphasis on sectoral location in the analysis of technical systems. One complex of environmental factors, including instrumentation, personnel, and work team structure, relates to the immediate research locale of the performer and may be termed the *organizational context*. Organizational context may be construed as consisting both in factors conducive to the production of work with certain technical characteristics and as factors leading to visibility within the technical system. Fiscal resources, equipment, support personnel, and prestige are all elements which are properties of the organizations within which scientists work. This dimension may be seen as the organizational environment for technical work--the more favorable the organizational context, the greater we would expect a researcher's contribution to be.

To what extent do organizational context and network centrality[5] affect scientific performance? Of course, to establish causality would require a study performed at two points in time. The most we can do with the present data is to establish an association between these dimensions and speculate on the reasons for it. However, as a first step toward the analysis of performance in technical systems, it is a reasonable one.

We ask, then, whether both organizational context and network centrality are associated with each of the two basic measures of performance and whether they are associated independently of one another. It could be that both high levels of performance and high levels of contact are due to position within certain research centers. This is particularly plausible for organizations with coordinating functions such as national laboratories which have large research contracts and hold integration meetings for designated research areas. Alternatively, a central position in the social network of the system might lead to both high levels of performance and, through organizational mobility, a position at one of the wealthier and more prestigious organizations in the field. Therefore, we must control for the effects of organizational context in establishing a link between centrality and performance and vice versa.

In formulating a hypothesis on the relationship among centrality, context, and performance we must take account of the possibility, raised earlier, that nominated and rated contribution are indicators of two distinct concepts. Nominated contribution seemed to be a measure of the extent to which individuals are defined as innovators by the system as a whole, while rated contribution is a better measure of quality as perceived by knowledgeable peers. If nominated contribution does reflect a social definition, then its primary determinant should be the position of an individual in the network of relations constituting the system, the degree to which he/she is connected to other individuals in the field. Organizational context should not affect nominated contribution directly, although it may be somewhat associated with it by virtue of the fact that those who are central to the network tend to be located in favorable organizational contexts. If rated contribution reflects knowledgeable evaluations of research output, then it should be affected both by network position (since social relationships influence both task performance and assessments) *and* organizational context (given the resources and visibility associated therewith). We predict, then, that:

> (6) Rated contribution will be positively related to both organizational context and the degree of connectedness to the system, while nominated contribution will only be related to connectedness.

Measures of nominated and rated performance, as well as total contact, or connectedness, with the system have already been described. Organizational context is indicated in this study by the *rated contribution of the organization* to which each researcher belongs. It is "contextual" in the sense that the value for this variable will be the same for each individual in an organization. Like the measure of rated performance for individuals, organizational context is constructed from individual ratings of each organization in the sample. After completing a series of questions on the frequency of contact and exchanges in the respondent's dealings with each organization, he/she was asked, "For each organization with whose work you are familiar, please evaluate its overall contribution to important new developments in the field within the past three years." Ratings ranging from one to four were then averaged for all respondents to create a rated quality score for each of the organizations in the sample. This score was then mapped onto the individual level file to create a contextual measure for each respondent. Organizational quality is taken here as a summary measure of the organizational context within which the researcher works. Hypothesis (6) predicts that the more favorable the organizational context, the higher the level of rated performance for the individual researcher.[6]

The results of a multiple regression analysis are presented in Exhibit 5.1. Using rated and nominated contribution as dependent variables, the effects of connectedness and organizational context were estimated.[7] The most important figures for our purposes are the partial correlations which show the net associations of each independent factor on the dependent variable, or the amount of variance explained by that factor alone.

The findings in Exhibit 5.1 support the hypothesis with one important exception. For nuclear waste, rated contribution is not associated with connectedness, but it is significantly associated with organizational context. Controlling for context reduces the correlation of contact and rated contribution from .19 to .09, contrary to expectations. However, controlling for contact only reduces the correlation of context and rated contribution from .27 to .22. The relative magnitude of these effects is reversed for nominated contribution. Here contact is significantly associated with nominated con-

Exhibit 5.1 Predicting Performance with Network Contacts and Organizational Context

NUCLEAR WASTE

	Dependent Variable:			
	Rated Contribution[1] (n=107)		Nominated Contribution[2] (n=112)	
	Simple r	Partial r	Simple r	Partial r
Total Contact[3]	.19	.09	.59	.54*
Organizational Context[4]	.27	.22*	.31	.12

Variance Explained	.08 (p=.01)		.36 (p <.001)	

PHOTOVOLTAICS

	Dependent Variable:			
	Rated Contribution (n=122)		Nominated Contribution (n=130)	
	Simple r	Partial r	Simple r	Partial r
Total Contact	.45	.35*	.59	.54*
Organizational Context	.39	.24*	.28	.05

Variance Explained	.25 (p <.001)		.35 (p <.001)	

*Significant at .05 level or better.

[1]Rated contribution is measured as the average of all quality ratings received by an individual *excluding* those within the same organization.

[2]Count of nominations received by an individual, reexpressed as the negative reciprocal root.

[3]Total number of linkages reported by others, reexpressed as a logarithm (base 10).

[4]Average quality rating received by the organization to which respondent belongs.

tribution and also accounts for most of the simple bivariate effect of context. The second part of the table shows the same result for photovoltaics with the difference that contact is also associated with rated contribution, as predicted.

In both systems, then, nominated contribution is significantly related to the extent of connectedness, independent of organizational context, and context itself has no effect. About a third of the variance may be explained by the centrality of the researcher in a network of social relationships within the system. This finding supports the notion that nominated contribution reflects the process of defining a researcher as innovative, since such definitions are social constructions which circulate within research networks apart from the judgments of research on which the reputations are initially based. We saw that only two-thirds of all nominations are given by those who claim familiarity with the work of the nominee. Individuals with many direct ties to others in the system are more readily defined as innovators than those with fewer ties because of their visibility and because of the greater opportunity to influence opinions.

Such personal influence is likely to be even more effective for nonresearchers and researchers who are not directly familiar with their work. Too, attributions of innovativeness are more readily accepted by those who have had personal contact with an individual. The end result is that a label of "innovator" is affixed to certain researchers such that even researchers in unrelated areas can identify them. It is highly unlikely that innovative labels are unrelated to the actual technical performance of individuals. What is suggested is that a social influence process stimulates and maintains a reputation once an evaluation of performance has been made in certain central positions within the network.[8]

The other finding in Exhibit 5.1 is somewhat of an anomaly. As expected, organizational context is positively related to rated contribution for both systems, implying that researchers benefit from affiliation with certain organizations. What is not clear is whether structural factors, resource investments, or institutional prestige is responsible for this association. Contrary to hypothesis (6), total contact is not related to rated contribution in the nuclear waste system. The small positive relationship between the degree of linkage to the system and rated quality of work was apparently due largely to the location of those with relatively high perceived contributions in favorable organizational contexts. Since this finding runs counter to results from virtually all previous studies of scientific productivity and seems counterintuitive, an explanation is called for. It may be that the measure of total connectedness to the system is simply not discriminating enough, representing a count of relationships with individuals in all sectors of the system. It is to this possibility that we now turn.

Linkages with Specific Sectors

The analysis in Chapter Four showed importance of sector in patterning the distribution of social associations within both nuclear waste and photovoltaic systems. It may be that the explanation of performance in technical systems has something to do with the *sectoral* distribution of a researcher's social relationships. That is, it might be important to know whether a researcher has contacts with individuals in specific sectors, rather than his total number of contacts with all sectors, because of the distinctive functions of sectors in technical systems.

Social linkages are not all of equal value for performance in technical systems. It is proposed that the value of contacts depends on the sectoral location of the actor contacted and the nature of the technical system. The sectoral location of the actor contacted is important because of three mechanisms which might influence performance. First, some actors are *resource distributors* (in particular those in the administrative component) or potential sources of research support in the form of contracts and grants. Such contacts could result in the acquisition of resources which directly affect performance. Second, some actors are *opinion leaders* (actors who are well connected within the system and whose judgments have a high degree of credibility) who might be expected to influence other actors in the system regarding the evaluation of innovativeness within the system. Finally, some actors are *gatekeepers*, or sources of technical information and ideas (Allen, 1977). Of course, a given individual may serve all these functions or none of them. However, we would expect resource distributors to be concentrated in the government sector, with gatekeepers and opinion leaders concentrated in the research sectors. The nature of the technical system--whether it is a public or private technology--is important, as we have argued throughout, in specifying the sectors most likely to provide information, influence, and resources.

The distinctive interorganizational structure of the nuclear waste system--with a relative monopoly of resources by the federal government and the majority of the research performed by national laboratories--leads to the prediction that these sectors will be more influential in determining performance evaluations than private firms and universities. The "public core," as we have termed it, controls and coordinates the system and is the locus of most of the communication ties. Further, this control is more effective given the absence of alternative sources of support from the private sector.

If the judgments of those who are most aware of innovation in the system and the direction in which the field is developing are most significant for a researcher's performance, then government and national laboratory contacts should be more valuable than those with other sectors. That is, the judgments of well-connected individuals are more important than the judgments of less well-connected individuals because of the former's influence in opinion formation. These individuals tend to be located in the public sector in nuclear waste. A second reason for the predicted importance of such contacts is the control by these actors over resources for research. Contacts are valuable sources of information regarding the shaping of research. Since national laboratory researchers do not compete for contracts it is likely that their relationships with governmental actors are more easily converted into resources than those in the private and academic sectors.

In photovoltaics researchers are less dependent on federal research funds. Private firms constitute the primary research sector and the role of national laboratories is significantly smaller than in nuclear waste. However, universities are an important sector in the communication structure of the system, receiving more communication linkages than other research sectors. It is likely, then, that contacts with these research sectors are more important in photovoltaics than contacts with government and national laboratories.

> (7) Linkages to the "public core" (government and national laboratories) will be related to nominated and rated performance more strongly than linkages to private and university sectors in the nuclear waste system; linkages to private firms and universities will be related to performance more strongly than linkages to government and laboratories in the photovoltaic system.

In order to test this hypothesis the measure of total contact is broken down by sector to create separate measures of contact with government, national labs, private firms and universities. Once again, the reports of others are used rather than self-reports of contact. For example, the variable "government contacts" is constructed by counting the number of frequent or infrequent contacts reported with a given researcher by individuals in the governmental sector. To assess the effects of specific sectoral contacts, rated and nominated performance are regressed on all four measures of contact simultaneously.[9]

Exhibit 5.2 indicates that rated contribution is associated with governmental contacts in the nuclear waste system, but not contacts with other sectors. Examination of the partial correlation coefficients shows that the effects of governmental contacts are only slightly reduced, controlling for contacts with research sectors. However, the small zero-order correlations of rated contribution with laboratory, industrial, and academic contacts are reduced to negligible levels when governmental contacts are controlled. Nominated

Exhibit 5.2 Predicting Performance with Sectoral Contacts[1]

NUCLEAR WASTE

	Dependent Variable:			
	Rated Contribution (n=107)		Nominated Contribution (n=112)	
	Simple r	Partial r	Simple r	Partial r
Contact with:				
Government	.30	.23*	.64	.22
National Labs	.16	-.06	.68	.38*
Private Firms	.12	.01	.42	.08
Universities	.18	-.01	.48	.09
Variance Explained	.09 (p=.04)		.52 (p <.001)	

PHOTOVOLTAICS

	Dependent Variable:			
	Rated Contribution (n=122)		Nominated Contribution (n=130)	
	Simple r	Partial r	Simple r	Partial r
Contact with:				
Government	.36	-.06	.53	-.04
National Labs	.39	-.02	.60	.12
Private Firms	.44	.16	.62	.15
Universities	.44	.16	.62	.20
Variance Explained	.22 (p <.001)		.43 (p <.001)	

[1]Only researchers included in the analysis. Independent variables are total number of linkages reported by others in specific sectors (government, labs, etc.).

*Significant at .05 level or better.

contribution is associated with both governmental and laboratory linkages, but not with private and university linkages, as hypothesized.

The results for the photovoltaic system are not so clear. Linkages to private and university researchers do somewhat better in predicting rated contribution than linkages with governmental and laboratory actors. The relative magnitudes of these sectoral contact effects are similar for nominated contribution. However, the differences between effects are not large and are only marginally significant, in spite of a highly significant overall fit. The reason for these small effects are relatively high correlations among the individual sectoral contact variables.10 Hence, it was decided to aggregate contacts for the research sector in photovoltaics in the remainder of the analysis.

Exhibit 5.3 (first column) presents the preferred models for nominated and rated performance, employing the best sectoral contact measures and organizational context. In the nuclear waste system 12% of the variance in rated contribution can be explained by contact with governmental actors and organizational context, with effects of similar magnitude.11 Thus, hypothesis (7), which argues that contacts with the public core are important to performance, is supported for contacts with government. For nominated contribution, 51% of the variance may be explained by contacts with government and national laboratories. Both effects are consistent with the "public core" hypothesis.

Preferred models for the photovoltaic system are also shown in Exhibit 5.3. Since research sector contacts are treated in aggregate fashion, we note that there is no reduction in predictive ability from Exhibit 5.1

Exhibit 5.3 Sectoral Differences in Determinants of Performance*

NUCLEAR WASTE

Dependent Variable: Rated Contribution

	All Researchers (n=107)	Labs (n=53)	Firms (n=31)	Univ. (n=23)
Organizational Context	.18	.16	.59	-.25
Government Contact	.22	.25	-.05	.49
Variance Explained	12%	12%	35%	25%

Dependent Variable: Nominated Contribution

	All Researchers (n=112)	Labs (n=57)	Firms (n=32)	Univ. (n=23)
Government Contacts	.32	.25	-.13	.73
Laboratory Contacts	.42	.41	.44	-.03
Variance Explained:	51%	49%	19%	64%

PHOTOVOLTAICS

Dependent Variable: Rated Contribution

	All Researchers (n=112)	Labs (n=33)	Firms (n=59)	Univ. (n=30)
Organizational Context	.25	.32	.20	.34
Research-sector Contacts	.37	.44	.28	.60
Variance Explained:	26%	40%	16%	45%

Dependent Variable: Nominated Contribution

	All Researchers (n=130)	Labs (n=36)	Firms (n=63)	Univ. (n=31)
Research-sector Contacts	.66	.64	.68	.75
Variance Explained:	43%	41%	46%	56%

*Coefficients are partial correlations, controlling for the other variable in the model. In column 1, all researchers are included, in columns 2-4, only researchers in specific sectors. (See exhibits 5.1 and 5.2 for construction of variables.)

where government contacts were included as well. The model for rated contribution is more successful than that in nuclear waste, explaining 26% of the variance, while that for nominated contribution is somewhat less successful, explaining 43% of the variance with only one variable.

Opinions and Resources

How, then, are these final models for performance to be interpreted? Nominated contribution reflects a social labelling process, whereby an individual is defined as an innovator. In nuclear waste, such definitions tend to be established by those in government and national laboratories; in photovoltaics, by those in the research sectors. Since social associations are a means of establishing visibility for one's work, we may want to say they affect definitions of innovativeness through a twofold influence process akin to the "two-step flow" process often held to characterize opinion formation (Lazarsfeld, et al., 1948). First, direct associations increase the likelihood that assessments of innovativeness will be favorable. Second, the ascription of innovativeness circulates within the system and increases the social standing of researchers who are characterized as innovative even among those not familiar with their work. This sort of cumulative advantage is based on the social associations maintained by researchers and appears to be unaffected by their organizational environment.

Rated contribution, on the other hand, is an evaluation by knowledgeable peers of the significance of a researcher's work. In nuclear waste, linkages with government are associated with high ratings, while in photovoltaics, high ratings are associated with research-sector contact. It seems likely that the mechanisms of resource distribution, opinion leadership, and gatekeeping explain these findings.

Governmental contact implies access to the sources of support and research resources within the nuclear waste system. For laboratories, which perform the lion's share of research in the system, this is particularly important. Since there is a relative absence of outside sources of research funds in the system, close relationships with governmental actors are critical for the establishment of research programs. The competitive advantage to be gained by maintaining such relations is large. One Department of Energy manager spoke of the difficulty of getting relevant proposals from university researchers, contrasting them with national laboratories:

> Their bread and butter depends on identifying hot areas and convincing us of their importance...the labs have an advantage given their formal channel of information flow. The directors know me and it's easier to talk to one person than to thousands of individual university researchers.

This is not to suggest that contracts are let based on personal friendships alone, though the trust developed over time is a form of insurance for program managers that contracts are well placed. Personal interactions with program managers involve the flow of information regarding federal program objectives and needs, such that research strategies can be developed.

Conversely, program managers call on trusted researchers to fulfill immediate research needs rather than send out Requests for Proposals, a costly and time-consuming process:

> It's easy to place contracts at national labs.... They send me a proposal, I do the paperwork and send them the money in about three weeks....You know the national labs and give it to who you think will do it best....There's a lack of competition.--lead center program manager

> The reason for using national labs is that if you use contractors there's a much longer delay in start up, given the need to advertise and get competitive bids. National labs have about a one month delay. Contracting out takes about nine months at least.--regulatory agency officer

It is not surprising, then, that relationships with government are associated with evaluated performance, given the potential access to resources which such contact implies.

In photovoltaics linkages with government are not associated with performance, but rather linkages with other researchers. It is doubtful that research contacts are often a means of resource acquisition, at least in the same sense that governmental contacts are. Given the relatively large private expenditures on photovoltaic R&D, actors within this system are not dependent on government to the same degree as in nuclear waste. Of course, private-sector actors are in a much better position to receive these resources from their own firms, but contracting relationships with other firms and research organizations are not unknown. The existence of alternative sources of research support reduces dependence on government and simultaneously the power of government in influencing agenda setting and research priorities in photovoltaics. Private firms are not unaffected by governmental funding patterns, but they remain autonomous to a significant degree with respect to their own research expenditures. Hence, government contacts are not associated with performance in photovoltaics.

Contact with researchers is more likely to affect evaluations through gatekeeping and opinion leadership processes than through resource distribution. Social relationships are channels for the transmission of ideas and information conducive to the production of high-quality research, the traditional explanation for the association between communication and performance in studies of R&D management. Secondarily, they are important as an opportunity to lobby for one's research contributions and directly influence colleagues regarding the significance of one's research. Finally, they are opportunities to perform favors--to give advice, information, samples, or assistance which benefits other researchers and encourages positive evaluations. The exchange of favors for positive evaluations is a direct benefit of social interaction. Evaluations circulate within the assessor's own personal network, leading to secondary evaluations.

Sectoral Differences in Determinants of Performance

Although the role of sector has been examined in terms of its importance for resources and reputation building, we have yet to consider whether the determinants of performance are similar for scientists in different sectoral locations. That is, the sectoral *linkages* of a researcher have been considered, but not his own sectoral *affiliation*. The preferred models in Exhibit 5.3 are estimated separately for each research sector in columns 2-4 (national laboratories, private firms, and universities).[12] Each sheds light on two questions. First, to what extent do the models which predict performance in the system as a whole also fit individual sectors? Second, do organizational context and contacts with government or research sectors affect performance within specific sectors?[13]

Exhibit 5.3 shows that, in the main, the photovoltaic system is more consistent in terms of the magnitude of effects by sector. Organizational context affects rated contribution in each sector, controlling for research contacts, in roughly the same degree as it does for all researchers. Research-sector contacts are more strongly associated with rated contribution for university and laboratory researchers than private researchers.

In nuclear waste there are larger differences by sector, underscoring the fact that it is a more heterogeneous and less integrated system, and arguing for distinctive determinants of performance for each sector. Organizational context affects rated contribution much more in the private sector than in national laboratories, and the effect is actually negative for universities. However, for contacts with government the case is quite the opposite. Here, private researchers do not profit at all from governmental contacts, while the association is twice as strong for university researchers as for the sample as a whole.

When the proportion of variance explained is considered, both models display a striking fact: performance in the primary research sectors is not nearly as well accounted for as performance in peripheral sectors. In nuclear waste, only 12% of the variance in laboratory performance is explained, while 35% and 25% is explained for private and academic sectors respectively. In photovoltaics only 16% of the variance in private-sector performance is explained with this preferred model as compared with 40% of the variance in laboratory performance and 45% of the variance in university performance. It behooves us to try to do better--a task we attempt in the final section of the chapter.

The anomalous effects for private and university researchers in nuclear waste do not readily admit of explanation. Why should researchers who come from less favorable university contexts be rated more highly themselves?[14] A partial explanation may lie in the referent of organizational context ratings, which is affected by the cognitive complexity of nuclear waste. Whereas in photovoltaics academics are likely to be located in departments of electrical engineering or physics, in nuclear waste they may be in geology, chemical or civil engineering, health science, chemistry or several other disciplines. Thus, ratings of organizations are less likely to reflect the immediate organizational environment of university respondents in this field.

The absence of an effect of governmental linkages for the private sector may be due to the institutional strains which characterize the relationship between private and government sectors in nuclear waste. The private sector is, of course, adversely affected by the lack of an effective waste disposal system in the United States and has tended to criticize the halting and changeable governmental policies for waste management. Several of these firms are reactor manufacturers and others are involved in the nuclear power industry in various capacities. It may be that interactions between government program managers and privately employed scientists are affected by these institutional relations; as a result private researchers experience more difficulty in converting governmental contacts into research resources, partly because program managers wish to avoid the appearance of conflicts of interest. Such an opinion was voiced by a researcher from a large private firm:

> Within the nuclear waste community there is a concern over conflict of interests. In giving contracts, private enterprise may not be able to do credible work because they may be too anxious to get things going and move on to the next stage of development....National labs have a distinct advantage in relation to government. They think private companies are biased.

Finally, university researchers seem to benefit more than other sectors from contacts with government. Although the sample size is small, we may speculate that this is also related to the legitimizing function of university researchers in the nuclear waste system. Social associations with university researchers are more valuable for program managers than those with laboratories. The latter are easy to acquire, numerous, and replaceable. University researchers are more difficult to interest in radioactive waste problems due to their relatively applied nature, yet they are quite critical for the success of the program. This was explicitly recognized by several informants, including one DOE program manager:

> We're trying to increase the number of university researchers [in our program]. Until now most research has been done at the national labs.

Academics may benefit from this special status through easier access to resources.

Performance in Primary Research Sectors

Sectoral participation has been employed throughout the analysis as a key dimension of technical systems. In Chapter Three we defined the *primary research sector* of a system as that research sector which accounts for most of the research activities in a system, as measured by the relative share of publications it produces. For this reason, the explanation of performance in primary research sectors is a special concern.

Exhibit 5.3 showed that the preferred models for rated contribution explained only 12%-16% of the variation in primary sector research performance. In attempting to improve this figure no excuses are offered for the plainly inductive approach followed. After presenting quantitative results of an exploratory search for correlates of rated contribution, case studies of researchers in radioactive waste and photovoltaics are employed to bring out characteristic features of innovators in the two fields.

The effort to increase the predictive value of the models for performance in nuclear waste laboratories and photovoltaic firms involved a reexamination of sectoral linkages, then the addition of variables which had shown promise in an analysis of organizational context.[15] Exhibit 5.4 shows the results of the regression of performance on four variables for nuclear waste laboratories. This model explains 36% of the variance in rated performance using organizational context and governmental contacts (as in the preferred model) and adding measures of diversity and size. Two findings are important to note here. First, the larger the organization (number of personnel working on radioactive waste) and the more sectors contacted, the

Exhibit 5.4 Predicting Performance in Nuclear Waste Laboratories and Photovoltaic Firms

NUCLEAR WASTE LABORATORIES

Dependent Variable:	Rated Contribution (n=47)	Variance Explained = 36% (p=.001)		
	b Coefficient (standard error)	Standardized Coefficient	Simple r	Partial r
Organizational Context [1]	.79 (.41)	.28	.21	.29
Sectoral Diversity [2]	-.25 (.09)	-.48*	-.04	-.39
Government Contact	.19 (.05)	.66*	.30	.48
Size [3]	-.25 (.08)	-.42*	-.21	-.41

PHOTOVOLTAIC FIRMS

Dependent Variable: Rated Contribution (n=55) Variance Explained = 25% (p=.006)

	b Coefficient (standard error)	Standardized Coefficient	Simple r	Partial r
Organizational Context [1]	.42 (.42)	.14	.33	.14
Autonomy [4]	.07 (.13)	.08	.22	.08
University Contact	.05 (.02)	.33*	.42	.33
Size [3]	-.16 (.12)	-.17	-.25	-.18

*Significant at .05 level or better.

[1] Average performance rating received by organization to which researcher belongs.

[2] Count of sectors reporting contact with a respondent (includes government, national laboratory, private, academic, public interest, and policy sectors). The range is from 0 to 6.

[3] Respondent's report of the number of professional personnel working in the field at his/her organization.

[4] Agreement/disagreement with the statement "I have a lot of freedom to select my own research problems"(four-point scale).

lower the rated performance. Second, the effects of organizational context and contact with the governmental sector are substantially larger after controlling for size and diversity. That is, their effects were masked to some extent by the omitted variables in the earlier model.

The effects of size should not be understood as referring to the absolute size of the national laboratory, but rather the size of the effort in nuclear waste. The negative coefficient indicates that, net of other variables, those whose research is conducted in contexts with large numbers of radioactive waste researchers are rated less highly than those in less populated contexts. It might seem reasonable to interpret this as providing support for the notion of the rigidity and lack of creativity in large, bureaucratized research organizations, a frequent complaint about national laboratories. As one lead center manager put it:

> There's too much bureaucracy in the labs... too many obstacles...[they're not] inferior quality scientists, just top heavy in management.

However, radioactive waste research is more concerned with the systematic application of knowledge than with producing basic knowledge. Since large research laboratories are usually held to be better at exploitation of innovations than in their creation there is another plausible explanation. The multitudes of research workers at the larger laboratories make it more difficult to single out the contributions of individuals from the vast quantities of output known to be produced by the laboratory. An individual researcher may produce work of very high quality, but lack visibility in the context of Sandia or Oak

Ridge national laboratories. Another individual may simply acquire a larger share of the credit due to the greater visibility of a smaller laboratory.

Sectoral diversity is measured as a count of the number of different sectors reporting contact with the researcher. The negative coefficient may be interpreted in the light of the political controversy surrounding nuclear waste research. Besides contacts with government and laboratories, nuclear waste researchers may have linkages to other research sectors (universities, private firms) or nonresearch sectors (public-interest groups, regulatory agencies, etc.). Some of these groups are known opponents of DOE policy and strive to reduce the credibility of the program by utilizing technical uncertainty to advantage. Linkages to these sectors may be threatening to DOE managers. Although this is far from suggesting that tight circles of insiders and outsiders exist which completely structure social associations within the nuclear waste system, the finding here suggests that a set of contacts focused on the governmental sector is conducive to highly rated performance. Indeed, the standardized coefficient for governmental contacts increases from .26 to .66 when controlling for sectoral diversity and size.

In photovoltaics the private-sector performance model is estimated in Exhibit 5.4. The improvement in fit is not as large as the threefold improvement in radioactive waste, amounting to a 9% increase (from 16% to 25%) in explained variance. Here, university contact replaces research-sector contact as the principal carrier of network effects. However, individual terms are not significant in this model with the exception of university contact. Once again, size is inversely related to rated performance. Freedom in the selection of research problems is not significant, but its addition to the model results in an increment of 5% in the variance explained by the model as a whole. The effect of organizational context is diminished slightly in this equation.

In sum, it is possible to do better in predicting performance for the primary research sectors, but at some cost in terms of parsimony and confidence in the stability of the estimates. Given the small size of the sample, the results should be treated as suggestive rather than as evidence for any theory of primary-sector performance. In nuclear waste laboratories, "focused" contacts with government and the character of the immediate organizational environment appear to be the important predictors of performance. The character of this environment is shaped by the quality of the organizational output and the number of intraorganizational competitors for recognition. "Focused" contacts imply strong linkages with governmental actors but not diverse linkages throughout the system. In photovoltaics, some improvement in prediction was seen, but the model seems inadequate given the relatively small increment in variance explained with two additional terms. However, some evidence was provided for a decrease in perceived individual contribution based on a larger population of researchers within an organization.

Typical Innovators

We have seen improvement, but are far from a complete understanding of performance, in large part due to limitations of the data. In spite of the fact that we have only survey data and not depth interviews, it is possible that some additional insight may be gained by examining two surveys in a nonquantitative fashion, together with abstracts of the research produced.[16] Two researchers were selected: SE, a forty-year-old chemist from a large national laboratory, and GG, a thirty-eight-year-old electrical engineer at a private firm. Both are highly rated as innovators in their respective fields.

GG's work received sixty-four ratings and ten nominations as innovative. His firm does not produce solar cells at present, but does better than average research in photovoltaics. GG himself has worked in more than half of the photovoltaic specialties listed in the questionnaire: single-crystal silicon cells, gallium arsenide cells, amorphous silicon, polycrystalline cells, and concentrator systems. He has produced a number of review articles on the field in journals such as *Solar Energy* and *Chemical Technology*. Like most photovoltaic researchers, he presents papers at the IEEE Photovoltaic Specialists Conference held every eighteen months. Another publication outlet is in-house technical bulletins. GG reports complete freedom in the choice of research problems. Recent projects have been the development of a low-temperature spray process for antireflection coatings and work on doping in horizontal multijunction cells. With two other researchers in his firm he was granted a patent on a semiconductor heterostructure cell. In work on concentrator systems he has studied novel materials and devices and shown in an economic analysis that cell cost is much less important than cell/concentrator efficiency and concentrator cost for such systems. By his own estimate, 60% of his work is applied, the rest split between basic research and development. GG feels his main contributions to photovoltaics have been empirical, including new techniques and methods. He reports one book, five patents, and twenty-five articles in the field.

Partly because his firm does not produce solar cells GG has no contact with sales and marketing professionals, but he has three times as many contacts with other photovoltaic researchers than the average. Professionally active, he has served on editorial boards, advised public-interest groups, and attended five meetings in photovoltaics in the past three years, spending nearly a month traveling within the past year.

GG's firm is rated as having made above average contributions to the field, though there are only five researchers working in photovoltaics. GG began working there after graduating from Carnegie Mellon with a Ph.D. in electrical engineering in 1968 and has never worked for another firm. He began working in photovoltaics in 1970, the same year his firm became involved. Sixty percent of his research support comes from the company's own expenditures. The remainder comes from grants and contracts from NASA and the state energy office. Significantly, none of these funds come from the DOE or its lead centers. Most of GG's current work (80%) is in photovoltaics, consisting of four projects, three of which he directs. He is involved with two projects in other areas.

Besides research activities, over a third of his time is spent on administration, working closely with twenty-five other scientists and engineers and supervising three professionals and three technicians directly. In spite of his private-sector affiliation, he works with two graduate students.

A number of similarities characterize the profiles of GG and SE, who is employed by a national laboratory. SE has worked at the same location since graduate school at the

University of Minnesota. He, too, was on the ground floor of radioactive waste work at his organization in 1974 and spends most (70%), but not all of his time in the field. He currently works on four projects, directing three of these, and participating in one other project unrelated to nuclear waste. Like GG, he has complete freedom in the selection of problems and has worked in a diverse array of fields (seven of ten listed specialty areas) including ceramic waste forms, salt/waste interactions, transmutation, risk analysis, and radionuclide migration in groundwater.

SE's research output is primarily in the form of technical reports for DOE-funded contracts. These include reports on the interaction of radionuclides with argillite from a formation at a nearby nuclear reservation, a study of the use of titanates in decontaminating defense waste, and examination of the ability of geomedia around the Waste Isolation Pilot Project to retard migration of radioactive substances. He has also collaborated on an alternative to the standard method of calcifying and vitrifying liquid waste, using inorganic ion exchangers for solidification and incorporating it into ceramic forms. He considers all of this work to be applied research.

Although he has not been a journal editor or officer in a professional association, nor given advice to public-interest groups, SE attends DOE integration meetings. During the past three years he has attended an average of six meetings per year, including forty-five days traveling during the past year. He has more than the average number of contacts with government, and the majority of his research communication partners are from national laboratories. Interestingly, he views government officials as unwilling to listen to new ideas and perceives an "old boy network" which controls the distribution of research funds--perhaps because his own work focuses on ceramics, an alternative waste form.

There are several differences between SE and GG, rooted primarily in the nature of their organizational contexts. Of course, the laboratory is also an innovative organization, but it is large in terms of the effort devoted to radioactive waste.[17] SE is somewhat unusual in that he devotes only a small proportion of his time to administrative tasks (10%) and does not supervise anyone directly, although he works closely with a dozen scientists and engineers and an equal number of technicians. Significantly, all of his research support derives from the Department of Energy.

These characteristics lend some substance to the regression analyses and render them intelligible by providing similarities and contrasts between specific individuals who are perceived by their peers as innovative. The typical innovator in both fields is a mid-career Ph.D. with strong linkages to an innovative organization. He has begun to take on some administrative responsibilities in addition to his research tasks and spends several weeks traveling each year, attending professional meetings which increase his visibility outside the local organization. Free to select his research problems, he is still actively involved in collecting and analyzing data and generally works closely with a number of other scientists in his organization on several projects at once, but does not devote his research activities exclusively to this one field.

Differences between these innovators arise from the relationship between the type of organization they work for and the technical system within which their contributions are judged. In photovoltaics, where firms are willing to invest in R&D and the size of the total effort is small (in terms of the number of researchers), innovators are less dependent on government for research support. By contrast, innovators in radioactive waste are likely to be affiliated with large, national research laboratories and are entirely dependent on the administrative component of the system to support their efforts through contracts and program grants. Hence, contacts with government and lead center personnel are especially important. They are more likely to publish findings in government report form, less likely to use journal outlets, and less likely than researchers in photovoltaics to be involved in basic research.

Summary

These researcher profiles and the quantitative analyses reinforce the general argument that the administrative component constrains but also facilitates the research behavior of individual scientists. However, the influence of the administrative component depends on the nature of the technical system--in particular the interorganizational structure which emerges from the interaction of political and economic factors. In this chapter the principal finding is that system properties have consequences for research performance evaluations at the individual level and, hence, scientific standing within technical systems. Associations with federal program managers are related to high rated performance in systems where dependence on government for research support is not offset by significant alternatives elsewhere. In these systems, connectedness to the research sector does not seem important for performance.

Peer evaluations are also associated with perceptions of the organization within which the research is produced. Research contributions which come from organizations producing high-quality work tend to be more highly valued themselves--but whether this is due to technical characteristics of the output or a perceptual bias is not clear. Support for a more traditional notion of the scientific community was found in photovoltaics, where there is more diversity in the distribution of researchers across sectors, less dependence on government, and alternative sources of research support. Here, as in studies of scientific specialties and individual firms, performance is associated with connectedness to the research network.

Chapter 6

Organized Technology: Opportunities and Limitations

Two case studies in the energy field are hardly enough to provide evidence for any general theory of technical systems. They are provocative, nonetheless, for their representation of two rather extreme types of system, one oriented more toward the private sector, one toward the public. In this final chapter the broad outlines of the argument are reviewed along with its implications.

The increase in the complexity of technology, its conspicuous success in many spheres of economic and political life, and the growing willingness of decision makers to seek technological solutions to social problems have led to the emergence of large-scale technological enterprises in advanced industrial societies. The resources required for the support of such enterprises are beyond those which a single firm can provide, and their output is often not profitable, at least in the short term. In the majority of cases, the federal government acts to organize these entities, which have been termed *technical systems*. Their primary objective is the production of technological innovation.

Technical systems have a number of characteristics which distinguish them from communities of basic scientists and collectivities organized by the industrial firm, both of which they may encompass. They are relatively large, both in resource expenditures and the number of actors they employ. They are diverse in terms of the breadth of technical expertise, the kinds of occupations, and the variety of sectors included. They are formally organized, with a division of labor, hierarchy of authority, and a central administrative entity such as a federal program office. As contrasted with the process of innovation within the firm, technical systems face extraordinary problems of control and coordination, given the scope of their tasks and the relative autonomy of component organizations.

While there is general agreement on these features, there is no established typology of technical systems. The distinction between collective and private goods was adopted as a conceptual device to illuminate the principal differences between nuclear waste and photovoltaic technology, the two fields selected for exploratory research purposes. Nuclear waste disposal is a collective good, since the benefits of an uncontaminated ecosystem are not feasibly withheld from an individual. Photovoltaic cells are private goods, since their benefits accrue primarily to those who pay for their production.

Both fields emerged in the 1950s as a response to political and economic problems, yet there is nothing inherent in the technologies themselves which require their development in the service of collective or private values. The history of photovoltaics demonstrates that the use of technology is politically determined, with the shift from space to terrestrial applications reflecting the emergent priority of private energy development over space exploration, a collective good. Nuclear waste management shows that, ironically, the growth of a system can be stimulated by a focus on technical issues, at least where that is accompanied by a neglect of its social environment. Further, the controversial character of this field was fertile ground for the establishment of successful disciplinary claims for support. During the period of study both systems grew as part of the more general "supersystem" of energy research and development, though the trend is unlikely to continue in the near term.

The use of technology for public or private purposes affects the interorganizational structure articulated in its development. All technical systems have some form of central administration (by definition), and most are supported (or plagued) by the involvement of public, private, academic, and policy sectors. Two important dimensions of difference exist based on the sources of financial support and personnel available to the system. Where collective goods are at stake, as in nuclear waste disposal, the state operates a virtual monopoly over expenditures for research, development, and production. The responsibility to maintain national laboratories owned by federal agencies is a strong motivation for undertaking a significant amount of technical activity in-house. Where private goods are at issue, the private sector provides support for development or supplements state funds. Hence, we find differences between the two fields in the diversity of funding opportunities and the distribution of research output across sectors. Whereas national laboratories generate most of the output in nuclear waste, private firms perform this role in photovoltaics. Otherwise, the scale of the effort, the formalized planning and management techniques, and the decentralized organization are similar in both systems.

The integration of public and private sectors is accomplished by government in both systems, as shown by an analysis of communication among sectors. While governmental actors have more linkages to all other sectors, the structure of associations within the research sector is more uniform in photovoltaics, more concentrated in national laboratories in nuclear waste. However, in terms of exchange rather than simple communication, there is not much difference between the systems. The sectoral location of an individual is more important than the type of system in shaping the likelihood of resource transactions.

Finally, the systemic constraints on innovation at the level of the individual researcher were examined. There appear to be two types of innovativeness in technical systems. The first reflects social reputational factors and is affected by the number and kind of relationships maintained within the system. The second corresponds more closely to traditional conceptions of technical performance, based on evaluations by knowledgeable peers, and is influenced by both social and organizational factors.

The most important finding overall is that linkages with government are associated with performance for researchers in nuclear waste but not in photovoltaics. Further, in nuclear waste contacts with the research sector were *not* associated with performance, whereas such contacts are in photovoltaics. The best explanation for this result is the difference in interorganizational structures of the two systems.

It is a basic notion of exchange theory that the greater the number of alternative sources of a reward, the less the power of any one source in controlling the behavior of other social actors. Conversely, to the extent that one actor monopolizes a valued commodity, that actor has greater potential to influence behavior. This situation characterizes the nuclear waste system *at the organizational level*, where a central actor (the federal program office) has a relative monopoly over the resources for research. In photovoltaics the private sector provides an alternative source of research funds and offsets the dependency on government characteristic of nuclear waste research.

The association between government contacts and performance and, at the aggregate level, that between innovativeness of a subfield and its level of activity are consistent with the notion that the innovation process takes on a different character in systems concerned with collective goods due to the domination of the state. Where the lead program office holds a monopoly over research revenues, its ability to influence the system through the control of resources will be great, particularly where contracts are allocated on a noncompetitive basis. Program managers are likely to cultivate researchers who are experienced and consistent, whose organizational position and track record make them good bets to perform well and, importantly, who have very low probabilities of failure. Likewise, those researchers who can build and maintain linkages within the federal program offices (such as senior scientists in laboratories or universities) can be assured of an ample and continuing flow of resources and will hold a competitive advantage over those who are not well connected.

Evidence for this process is not apparent in the photovoltaic system, where contact *within the research sector*, but not with government, is associated with performance. This result supports the traditional view of science in the context of a new organizational form, the competitive technical system. Here, the greater the number of linkages to other researchers, the higher one's assessed contribution, owing to the greater visibility and information potential which such connectedness implies.

Understanding Science and Technology

It has been clear for some time that it is mistaken to view science as characterized by the lone researcher, dependent on small-scale experimental equipment purchased from his own pocket. Discussions of Big Science have recently begun to emphasize the team, the increasing complexity and cost of experimental apparatus, and the dependence on institutionalized facilities and support--the university, the industrial firm, the state. But contemporary science may only be different from classical science by virtue of its colonization of *all* available institutional sectors, with none left for further expansion (Mulkay, 1980). Doubtless behind such discussion is always the implicit question: how will such factors affect the process of scientific and technological development itself? Until now, there has been more speculation than evidence on this subject.

Jerome Ravetz crystallized much of the previous thought on the subject of the science-government relationship by suggesting that linkages with administrative actors are a new kind of scientific property:

> With the concentration of decision-making power to the investment agencies and their few advisers in each field, their estimate of a man is of more practical significance for his career, than that of some future diffuse consensus. Hence the location of a successful scientist's property tends to shift from his

> published results to his existing research contracts, and the personal contacts that will ensure their continuation. (1971: 46)

Ravetz himself takes these contacts to be important regardless of the systemic context within which they are embedded. Indeed, most popular writers continue to view science as an undifferentiated whole or a collection of disciplines, as if the researcher were a free-floating intellect whose selection of problems and methods depended primarily on his sense of what is important.

Of course, serious students have understood that educational and organizational constraints are paramount, utilizing such concepts as "research specialty," "problem area," or "technological field" to delineate intellectual and social boundaries and serve as macrounits of analysis. Efforts are generally made to classify such units, in the hope that such a scheme can identify units with similar features and predict phenomena of interest. In the present report, evidence has been presented that nuclear waste, but not photovoltaics, displays a pattern in the relationship between contacts and performance unlike that of basic scientific specialties. In essence we have tried to specify one context in which the Ravetz "hypothesis" holds true--that is, a field which is not (primarily) basic science, which involves a centrally administered technological program, a primary research sector consisting of national laboratories, and a federal monopoly on research funds. It has been contrasted with another applied field, similarly organized, but with a greater involvement of the private sector in both research and funding. In this field the expected relationship between contacts and performance is evident. Apparently, there are conditions which can prevent or retard the shift in the nature of scientific property and the growing importance of government.

In brief, the results of the study emphasize:

(1) that *not all communication* facilitates innovation among individual researchers in technical systems: it depends on the type of system and the type of organization to which one belongs.

(2) that *nonresearchers* (in particular, administrators and program managers, but also policymakers and public-interest advocates) are important actors in technical systems: their integration into technical networks and relationships with researchers warrant much further examination.

(3) that *sectoral relationships* may be studied in a quantitative, systematic fashion by focusing on the rate and type of social associations between individuals. Further, exchange relationships may reveal more consistency across systems than simple communication.

(4) that "industralized science" does indeed display some unique features, but they may be more unique to systems which rely on government than to those which involve private industry.

Research Managers and Types of Communication

According to Sayles and Chandler, the project manager in a technological system is like an organizational metronome, designed to keep a number of diverse elements responsive to a central beat (1971: 207). The supervisory challenge involved is quite different from that of most managers, for authority is by no means commensurate with responsibility. Rather than directing and controlling, the most relevant managerial behaviors are bargaining, coaching, and, occasionally, confrontation. Paradoxically, the larger and more complex the system, the more difficult it is to control through precise, systematic, rational techniques of management. Instead, "precise coordination must be attained by repetitive, personal, and diffuse administrative techniques" (Sayles and Chandler, 1971: 225). While there is little reason to dispute the need for negotiated personal influence in professional project management, these results provide some grounds for optimism regarding the potential for project manager influence in technological enterprises of a certain kind.

First of all, the assumption that communication *per se* fosters high technical performance has been called into question. At the level of the individual firm, Thomas Allen, Michael Tushman, and Denis Lee (1979) have shown that alternative patterns of communication are associated with performance in different phases of development. In the broader context of a technical system oriented toward a collective good, communication with researchers was shown not to affect performance. Without the photovoltaic findings one would surely be tempted to argue that communication with researchers in general is superfluous due to the cognitive complexity of technical systems. But there is a strong association for photovoltaic researchers between contacts with the research sector and technical performance. For private researchers, university contacts were particularly effective, owing to the proprietary nature of the field and the less restrictive character of university research.

For the project manager, then, it may be well to distinguish between communication to *innovate* and communication to *interface*. For the former, the informational value of the communication is most important: ideas and directions which stimulate creative responses to technical problems. For the latter, the coordination of the system is crucial: the elimination (or, as the case may be, production) of redundancy, the design of interdependent components, and the clarification of specifications for the product. This sort of communication will not benefit the individual in terms of the level of performance achieved, but is significant for maximum levels of system performance.

Where communication to innovate is at issue, the manager must consider two points. First, the organizational restrictions on communication characteristic of private firms leads to lower "return on investment" in this sector. That is, from a managerial standpoint a system might be thought of as a mechanism for the conversion of resources into information, which then flows through the system to produce innovation. However, certain channels within the system (those extending from private firms) have greater resistance, so that resources invested here do not produce as much "free-flowing" information as resources invested in academic organizations or national laboratories.

Second, some qualitative evidence indicates that informal relationships break down the barriers to the transmission of information, even among private researchers. It is well worth seeking a precise understanding of the conditions which maximize this effect, promoting information flow in the system as a whole.

The radioactive waste field is instructive for interface communication. Although there are basic scientific advances being supported by program funds, the problems established by federal and lead center managers are generally reducible to "normal" technical procedures:

> New developments and contributions are not the name of the game. It's not a new science...but we look for people who can get things done.--program manager

The skills required to perform most technical tasks lie within most researchers' range of competencies. The necessary background is either part of one's repertoire, acquired through education and experience, or easily acquired through publicly available technical reports or proximate colleagues.

If this is the case, interface communication may be the most important type of relationship to promote. The use of national laboratories is indicated because of their ability to support systematic, ongoing developmental efforts which require large numbers of technical personnel. There may be little need to promote communication for innovation (for example, contacts with researchers in diverse fields). If the success of the technical system does not depend on astonishing levels of creativity--as most of our informants seem to suggest--the development of cogent plans and effective managerial practices is all the more important.

The notion of a "public core" in the nuclear waste system and the association of performance and administrative contacts does recall the idea of a "nuclear club" or "old boy network," often held responsible for the defective managerial practices and unresponsiveness of the Atomic Energy Commission (Chapter Two). Informants in public-interest groups and universities were quick to point out the exclusivity of the group, its refusal to provide funds to critical researchers, its abrupt dismissal of findings potentially damaging to preferred technological options. It is not unreasonable to apply the standard sociological wisdom here. A high degree of solidarity within the group facilitates the exchange of information and favors, but reduces the impact of external associations and criticism. The effective manager may profit by recruiting opponents of the program, even at some cost in terms of the efficiency of development.

The findings of this study do suggest that project managers have a role in the creation of reputations through their allocation of resources. Although this seems to be true only under very restrictive conditions, they are the conditions which are most likely to obtain where influence is most needed, namely in programs which are the exclusive responsibility of government.

The other side of the coin every manager knows--it is valuable to tie one's program to those who already *have* reputations, for they are sure bets. But the finding that nominated and evaluated performance are different points to a danger inherent in the usual practice of seeking innovators. Managers are most likely to use an analogue of the "nominational" approach in seeking proposals, members of panels, review committees, and other system functions. The problem with this is that actors uncovered through such means are likely to be those who are highly visible and recommended on grounds other than the technical merit of their work. The alternative--more arduous, costly, and defensible--is the targeted, comparative rating of contributions by knowledgeable peers. The model used by the National Academy of Sciences in the selection of study groups, first developing a pool of experts before selection, approximates such a two-stage method. A naming procedure, followed by a comparative rating, prevents reaching a premature consensus.

Policy Issues in Technical Systems

For policymakers, the differences between the nuclear waste and photovoltaic systems are important to consider. One response to the finding on communication and performance in nuclear waste is to seek ways of altering the dependency relationship between researchers and government. Interpreting the nuclear waste pattern as "deviant" and in need of change implies the advisability of checks on the process of monopolization of research capital by the core agency.

One approach, derived from the increasing diversity in federal agencies funding radioactive waste research, is to supplement, through legislative mandate or executive direction, a central program research budget with relatively equal confirmatory research budgets to be administered by different agencies. This would require more than the small proportion of nuclear waste research funds now administered by the Nuclear Regulatory Commission and the Environmental Protection Agency. However, it is not a promising approach for two reasons. First, accountability requirements make such large amounts highly vulnerable budget requests. If appropriated, pressures to coordinate programs would be difficult to resist. In this case the agency with superior technical expertise could be expected to dominate research planning, with the newcomer offices playing subordinate roles. This would undercut the benefits of autonomous, alternative support sources. Second, the career histories of most researchers with the relevant expertise are tied to the primary agency, to which loyalty and favors are owed. The pool of expertise is then restricted to those who are least likely to disagree with agency aims and initiatives. Of course, to the extent that (relatively) low levels of expertise are required to accommodate the research objectives this may not be a problem. Too, declining support for basic science has the latent benefit of increasing academic participation in technical systems and providing researchers with shorter histories (hence independence) with primary agencies.

Another possibility is to encourage the growth of private-sector research investments. Indications from the photovoltaic case point to the diminished influence of federal program managers on research performance and a facilitative pattern of contacts with the research community. The distinctive conditions for this state of affairs are related to the perceived profitability of solar cells. Can similar conditions be created in systems oriented to the production of public goods?

By definition, a public good is indivisible. Incentives must be legislated to motivate its provision by the private sector. The federal government defined its responsibility three decades ago as the promotion of nuclear energy, just as the early 1970s saw a commitment to solar energy. The promotion of nuclear power and the disposal of waste products deriving therefrom go hand in hand, yet the means for accomplishing them are somewhat conflicting. The former is taken to imply a relatively low public profile, emphasizing the speed and care with which the problem is being solved. An effort to involve the private sector in R&D would, on the other hand, imply an effort to convince company management of the wide scope of the problem and the profitability of developing new waste forms and containment devices which could be utilized. Just as firms which develop new pollution-control devices can market them for profit and are motivated to invest in pollution R&D, it is conceivable that a flexible regulatory approach could motivate investment in radioactive waste research. Nonetheless, the same uncertainties and policy shifts which have made firms reluctant to invest in reprocessing plants are present in radioactive waste research. The likelihood that actions of the federal government could assuage such concerns is low.

If private investment in nuclear waste research is to be stimulated, direct incentives must be provided. Tax credits could be used as the reward for research in the national interest. Firms could be guaranteed contracts for production of devices which meet certain specifications, again involving the regulatory agencies to a greater degree in the nuclear waste research system.

But whatever the viability of these options, they are predicated on the assumption that the pattern of dependency characteristic of the nuclear waste system is in some fashion detrimental to technological innovation, based on comparison with dissimilar units of analysis. This assumption may well be false. It is by no means clear that the nuclear waste system has performed less effectively than the photovoltaic system where *technical* issues are concerned (a comparison avoided with care throughout the analysis).

An intensive study of NASA management practices led Sayles and Chandler to conclude:

> The development group needs a working arrangement that will insulate it from its environment, and a monopoly or near monopoly of certain relationships is one way of achieving this goal. To get on with the job, the sponsoring agency is almost forced to make itself the central figure in a closely knit group of organizations, insulated from external pressures--from the environment--and therefore dependent upon the sponsor. (1971: 71)

Where high levels of coordination are more important than high levels of technical creativity--that is, where interface communication is at a premium--there are advantages to the kind of system represented by radioactive waste. The analysis of intra-sectoral relationships revealed that the primary research sector in radioactive waste (national laboratories) scored above the system average on both communication and exchange while the primary research sector in photovoltaics (private firms) scored below the average on both measures. While it is noncompetitive, a system of national laboratories is probably more effective in the transmission of information than a system of private firms. We are left with the paradox that the same inter-organizational conditions which promote the influence of managers and administrators also promote exchange and information flow which help to integrate the system. Conversely, those conditions which create a more "natural" relationship between communication and peer evaluations are those which reduce the level of information available to the system as a whole.

The role of the national laboratories in technical systems is in some ways the key to the process of innovation, at least from a policy viewpoint. Their proximity to federal agencies sets them apart from the community of academic scientists, as this and other studies have suggested (e.g., Sutton, 1977). Recently a report by the White House Science Council concluded that the nation can "no longer afford the luxury of isolating its Government laboratories" (*New York Times*, 16 July 1983), citing poor management, inability to attract top scientists, and low-quality work as problems with laboratories in a variety of fields. The dependence on government which creates close-knit circles of program managers and researchers has also been viewed as a source of nonresponsiveness due to the protected status it entails (OTA, 1980; Sayles and Chandler, 1971). Clearly, it would be desirable to separate the role of noncompetitive contractual arrangements from the monopoly of funds by the federal government in the technical innovation process.

Exploratory studies need not emphasize the usefulness of further research in conclusion, since it is built into their very conception. Whether the stress is on interpretation of evidence or hypothesis formulation, more questions will be raised than answered. This report has considered nuclear waste disposal and photovoltaic cell development as instances of *technical systems*, centrally administered networks of actors oriented toward the achievement of technological objectives. Such systems represent the most distinctive organizational forms in modern science and technology (Hannay and McGinn, 1980). They have been largely neglected as a subject for social scientific analysis, yet there are already textbooks on their management. Of course, the formal organization intrinsic to such systems ensures that people will be employed in their management before there are social scientists to poke around. But that will no longer do as an excuse.

Notes

Chapter 1
Technology and Technical Systems

1. Los Alamos alone employed 2,000 scientists and technical personnel in 1945 (Kevles, 1971).
2. The Oak Ridge installation used a total of 110,000 construction workers to build both production facilities and a town (Robinson, 1950).
3. Twenty private firms received "E" awards for their contributions.
4. The need for a sophisticated managerial plan is indicated by the complexity of the Apollo command module, which contained over two million functional parts.
5. Other writers have proposed such terms as "large-scale, public-private technological enterprises," "R and D systems," and "sociotechnical systems," but the term "technical system" seems to express the central idea as well as any of these.
6. This pattern is evident in all three systems discussed above.
7. The concern with innovativeness at the level of the individual scientist, the subject of Chapter Five, may seem out of place given all that has been said about the need to consider larger units in the organization of modern technology. But apart from the imperative for social scientists to examine the objects of the social world for their own sakes, there is good reason to think that these larger units constrain the behavior of individual scientists. It is here that the argument for their importance is strongest. If patterns at variance with those expected based on studies of academic and firm-based science are observed, our contention that technical systems warrant separate investigation will be supported.

Chapter 2
Collective Goods and Private Profits

1. The *cui bono* principle has been used in developing typologies of formal organizations (Blau and Scott, 1962).
2. Less than a year after the passage of the act, the Department of Energy announced that this congressional deadline would have to be pushed back at least three years and that a repository would not be in operation until at least three years after the 1998 deadline.
3. More recently the Reagan administration has reversed this policy again. However, at the time of this study the Carter policy was still in effect.
4. Clearly there is a need for an informed microanalysis of the use of "informed technical opinion" as an accounting device.
5. Just as the recent literature on firm-based innovation has begun to reject the dichotomy between market need and technological opportunity.
6. More recent estimates suggest under the most favorable conditions photovoltaic energy costs four times as much as conventional power.
7. This zone is crucial to the operation of the cell and is formed by the immediate and spontaneous migration of some excess electrons from phosphorus in the n-layer to the gaps or holes in the p-layer. The consequence is that the phosphorus atoms at this junction have lost electrons and possess a net positive charge, while the boron atoms possess a negative charge. There is, at this point, a barrier which filters out electrons moving at low energy levels, which means that the fewer high-speed electrons in the p-layer can cross over into the n-layer, but the many low-speed electrons in the n-layer cannot cross back. This causes a buildup of electrons in the n-silicon.
8. The nature of the data here precludes drawing any unambiguous conclusions, since the sample was not random and overrepresents elite members.
9. The reader may wish to refer to the second section of Chapter Four at this point, concerning the sample and design of the survey.

Chapter 3
Organization and Participation

1. One indicator suggesting that they are in roughly similar stages of development is the establishment of journals in both fields during the period of study (*Solar Cells; Radioactive Waste Management*).
2. One of the first instances of such delegation was the assignment by the air force of Ramo-Wooldrige Corporation for technical direction and systems engineering of the ballistic missile in 1954 (Kast and Rosenzweig, 1965).
3. Subsequently it was repositioned to report directly to the Secretary of Energy, reflecting the increasing policy significance of the field.
4. During the period of the study the program was "upgraded" from a branch to a division, but it is not clear that this represents any real increase in status, as both designations are four levels down in the hierarchy.
5. Further, under a special Agreement States program, authority to regulate certain types of nuclear waste is delegated to twenty-five states within their borders.
6. As a result of its geological expertise it has a critical role in both the selection and the licensing of repositories. Fears of a conflict of interest in this dual function have reminded some observers of problems in the old AEC.
7. A total of about 115 USGS researchers around the country worked on radioactive waste in 1980, with the largest concentration in Denver, Colorado. The headquarters are in Reston, Virginia, and the program is relatively decentralized, with a good deal of involvement by the program managers in the research.
8. This may well change if photovoltaics come to supply a significant share of our electrical energy.
9. At NASA there is a tradition of combining the project monitor role with the research role, which is separate in DOE management practice. Further, DOE tends to rely less on university capabilities.
10. These biannual meetings go back to the first uses of photovoltaics on NASA and air force satellites. Interestingly, informants in DOE and NASA disagreed on whether one purpose of the meetings was to reduce redundancy, indicating that indeed it is not.

11. Many classification difficulties are glossed over in the final results from this count, but other methods of classification would not change the finding of overall similarity of the two systems. One problem was the classification of national laboratories. Here and in Exhibit 3.5 NASA laboratories are classified as governmental agencies because they are owned and operated directly by the government.
12. The criteria for classification into large and small businesses are not stated in the report.
13. The figures for NWTS contractors do not reflect the actual distribution of research in the system by sector, as shown in Exhibit 3.2, but university researchers are overrepresented in peer review committees relative to either distribution.
14. Exceptions would include some celebrated defections of scientists at national laboratories.
15. Texas Instruments has established such an arrangement with DOE to develop a new photovoltaic technology.
16. Recent studies, including an unpublished report by the Office of Technology Assessment, offer this conclusion.
17. National laboratories do compete with each other for major research programs and program management activities. What is at stake here is the absence of competing proposals for specific projects.

Chapter 4
Communication and Intersectoral Relations

1. This project was funded under a grant from the Innovation Processes Section of the National Science Foundation (PRA-7920573) to the Princeton University Department of Sociology. Robert Wuthnow was the principal investigator.
2. Details concerning the design and methodology of the study may be found in Wuthnow, et al., 1982.
3. Respondents excluded their own organization here.
4. Higher rates of intrasectoral communication are largely responsible for this, since we have sampled a larger number of respondents from private firms in photovoltaics and from national labs in waste. These sampling differences are controlled for in the sectoral density models.

5. It would be erroneous to draw the conclusion, based on a comparison of these percentages, that photovoltaics is a more integrated field than radioactive waste. That much of the interaction in photovoltaics is simply due to the presence of the respondents' actual networks in this sample is revealed by a comparison of the percentages for organizational contact, measured in much the same way as individual contact. In this case, the names of organizations in the sample were substituted for the names of individuals, and the same question asked (minus the category "heard of, no contact"). In both fields respondents (by their own report this time) had contact with 43% of the listed organizations. A plausible explanation for the difference in individual level of contact is the larger number of researchers per organization in nuclear waste (see Chapter Three), leading to a reduced likelihood that the respondents' actual contacts were included in our sample.
6. This relatively high figure reflects the elite nature of the sample. The difference of ten percentage points between nuclear waste and photovoltaics is due to the greater number of policy actors in the sample for this field.
7. Doubtless the process is occasionally reversed as well. Individuals who are friends or professional colleagues may begin working on similar problems, or one may be drawn into the field by another. Here, as in the development of basic scientific specialties, institutionalized relations may emerge from initially informal connections.
8. All variables are treated as dichotomous in the table. Contact with respondents and researchers was dichotomized at the median. Standard survey items were simply collapsed from four categories into two.
9. Because of the small sample size, relationships must be fairly strong to achieve significance at the .05 level.
10. Tom Allen has examined rates of association for researchers across sectors in Ireland. His findings showed the closest associations between universities and the public sector, with low rates of contact between universities and private industry (1973).
11. These "policy" organizations are public or quasi-public bodies such as congressional committees, regu-

latory agencies, and national scientific advisory bodies (1973).

12. Row marginals represent the proportion of possible contacts *reported* by a given sector to all sectors. Column marginals are preferable as an indicator of overall contact for the same reason that the reports of others are preferable to one's own report of one's communication ties.
13. Diagonal cell values in Exhibits 4.5 and 4.6 have been adjusted to remove intraorganizational ties. Since most intraorganizational linkages actually occur, this lowers the diagonal values slightly, but almost uniformly, so that their relations remain the same.
14. In these models, both row and column marginals provide consistent rankings, though this is not always the case for such matrices.
15. Another indicator of status, using the same approach, is to compare *specific* intersectoral relations on reported ties in each direction. University respondents again receive more choices than they give in all comparisons.
16. This may not be true of academic researchers under contract to the Department of Defense.
17. The figure of .125 is based on eight relationships, too few to place confidence in as an accurate estimate.

Chapter 5
Research Performance in Technical Systems

1. Sometimes the argument is made that marginal individuals are important to innovation, especially via their access to new methods which can be imported into a field, but this does not controvert the fact that individuals who are centrally situated in a system are involved in a larger number of innovations than those who are peripheral.
2. "Innovativeness" and "quality of work" are used as synonyms of "performance" throughout the text.
3. Only researchers have been included in the calculations for these correlations.
4. These correlations are between raw (un-reexpressed) batches of numbers. In all of the analyses to follow, nominated performance is reexpressed due to extreme skewness in the distribution as a negative reciprocal root (Tukey, 1977). When so reexpressed, the correla-

tion with rated performance is .32 in nuclear waste and .54 in photovoltaics.

5. "Network centrality," "connectedness," and "contacts" are used synonymously.
6. "Organizational context" represents a group of variables. It is the best available indicator of non-network, environmental factors in the absence of a direct measure of resources. Nor is it necessarily inferior to financial resources as an indicator of "resources" since financial resources are clearly mediated by the organizational context. Of course, due to differentials in resource distribution and the complexity of the organizational environment, it would be a mistake to take the overall level of resources as an indicator for resources in any given field. (The facilities for doing research in photosynthesis may not be relevant to research in photovoltaics.) Therefore, the indicator used here refers to contributions to the field of photovoltaics/radioactive waste.
7. B's and beta coefficients are not shown.
8. The Matthew Effect in science refers to the differential recognition accorded to specific contributions made by elite scientists (Merton, 1973). "Accumulative advantage" is the process by which resources accrue to those scientists who make contributions early in their careers, leading to increasing inequality of publications and citations in a cohort over time (Allison and Stewart, 1974). All of these notions depend theoretically on the notion of some kind of label or social definition based initially on task performance and operating to increase visibility for the performer.
9. Although we found organizational context to be significantly related to rated contribution for both systems, this variable is excluded from the present analyses due to the small number of cases involved. After selecting sectoral contact variables, organizational context is reintroduced in Exhibits 5.3 and 5.4.
10. The Pearson correlations among sectoral contact variables show multicollinearity to be a significant problem in the photovoltaic data, given correlations ranging from .65 to .85. Among the research sector, correlations range from .75 to .84. In nuclear waste the correlations range from .30 to .70, and the ef-

fects of individual regression coefficients are either clearly significant or "insignificant."

11. Partial correlation coefficients are virtually identical to beta coefficients here.
12. Although it might be preferable to include a detailed analysis of each system by sector, the primary constraint is the small size of the sample. This has been a factor in the decision to be less concerned with statistical significance than with an overall pattern of results and in the decision to fit models with few variables. It applies *a fortiori* to an analysis by sector, where the number of cases ranges from twenty-three for the universities in nuclear waste to sixty in the private sector in photovoltaics. Because of this constraint, the approach used is to fit the preferred models for the system as a whole to each sector singly. Hypotheses will not be tested, nor significance levels used.
13. For the remainder of the discussion we focus on rated contribution as the dependent variable of interest.
14. It is worth noting that the negative effect is only significant at the .25 level (n=23).
15. The first step entailed a regression of rated contribution on the sectoral contact measures as in Exhibits 5.2 and 5.3. Although there was not much difference in fit between these models and the preferred models in the preceding section, there were differences in the role of contacts with specific sectors. For nuclear waste laboratories, the effects of government contacts were positive, but those for contacts with laboratories, private firms, and universities were negative. For privately employed photovoltaic researchers, only university contacts have any effect (partial correlation = .18). Not surprisingly, private-sector contacts are not associated even slightly with performance for private-sector researchers (partial correlation = -.01). Additional variables were added based on an analysis which sought to specify the effect of organizational context (not shown). Although these variables did not reduce the effects of context, as hoped, they were marginally significant in some cases. It was reasoned that they might add some explanatory power to models of performance in primary research sectors. Models in this

case were selected based on the highest R-square. It appeared that some dimensions, while not significant statistically, were important to control due to their suppressing effects on other dimensions.

16. Cases were selected from the primary research sectors in each system to represent typical innovators. Researchers from the primary sectors were preselected at one standard deviation or more above the sample mean on rated contribution. Thus, eight nuclear waste laboratory researchers and fourteen private photovoltaic researchers were initially taken. Next, scores on the variables which proved most important in predicting performance (Exhibit 5.4) were examined. Again, using the mean of the sample as a criterion, a selection was made in terms of the consistency of scores with the model. For example, if the size coefficient was negative in predicting rated contribution, a researcher was considered typical if his organization was below the mean in size.
17. SE did not report the number of individuals in total, although he claims contact with twenty-five per week who also work in the field.

Bibliography

Aldrich, Howard. *Organizations and environments.* Englewood Cliffs, N.J.: Prentice-Hall, 1979.

Allen, Thomas. "Institutional roles in technology transfer: a diagnosis of the situation in one small country." *R&D Management* 4 (1973): 41-51.

_____. *Managing the flow of technology.* Cambridge: MIT Press, 1977.

Allen, Thomas, Michael Tushman, and Denis Lee. "Technology transfer as a function of position in the spectrum from research through development to technical services." *Academy of Management Journal* 22 (1979): 694-708.

Allison, Paul, and John Stewart. "Productivity differences among scientists: evidence for accumulative advantage." *American Sociological Review* 39 (1974): 596-606.

American Physical Society. *Principal conclusions of the APS study group on solar photovoltaic energy conversion.* New York: American Physical Society, 1979.

Backus, Charles, ed. *Solar cells.* New York: IEEE Press, 1976.

Bell, Daniel. *The coming of post-industrial society.* New York: Basic, 1973.

Beniger, James R. "Using the principle of least interest to derive a dominance hierarchy from interaction or exchange data." *Proceedings of the American Statistical Association,* pp. 740-45 in section on survey research methods, 1980.

Beniger, James, Wesley Shrum, Thomas Ash, and Jerome Lutin. "Designing a survey of information and favor exchange among state, county, and municipal levels of government." *Proceedings of the American*

Statistical Association, pp. 182-87 in section on survey research methods, 1979.

Berkowitz, S. D. *An introduction to structural analysis: the network approach to social research.* Toronto: Butterworths, 1982.

Blankenship, L. Vaughn. "Management, politics, science: a nonseparable system." *Research Policy* 3 (1974): 244-57.

Blau, Peter. *Inequality and heterogeneity.* New York: Free Press, 1977.

Blau, Peter, and W. Richard Scott. *Formal organizations: a comparative approach.* San Francisco: Chandler, 1962.

Blomeke, John O., Jere P. Nichols, and William C. McClain. "Managing radioactive waste." *Physics Today,* August 1973, pp. 36-42.

Boffey, Phillip M. *The brain bank of America: an inquiry into the politics of science.* New York: McGraw-Hill, 1975.

Breiger, Ronald. "Career attributes and network structure: a blockmodel study of a biomedical research specialty." *American Sociological Review* 41 (1976): 117-35.

Brooks, Harvey. "The public concern in radioactive waste management." In *Proceedings of the International Symposium on the Management of Wastes from the LWR Fuel Cycle* (Denver, July 1976), pp. 52-60, Washington, D.C.: ERDA, 1976.

Bupp, Irvine C. "The nuclear stalemate." In *Energy future,* edited by Robert Stobaugh and Daniel Yergin, pp. 108-35. New York: Random House, 1979.

Busch, Lawrence, and William Lacy. *Science, agriculture, and the politics of research.* Boulder, Colorado: Westview Press, 1983.

Carter, Luther. "Radioactive wastes: some urgent unfinished business." *Science,* 18 February 1977, 661-66, 704.

———. "Nuclear wastes: the science of geologic disposal seen as weak." *Science,* 9 June 1978, 1135-37.

———. "Academy squabbles over radwaste report." *Science,* 20 July 1979, 287-89.

Chalmers, Bruce. "The photovoltaic generation of electricity." *Scientific American,* November 1976, pp. 34-43.

Chubin, Daryl. "State of the field: the conceptualization of scientific specialties." *Sociological Quarterly* 17 (1976): 448-76.

Cohen, Bernard. "The disposal of radioactive wastes from fission reactors." *Scientific American* 236 (1977): 21-31.

Collins, Randall. "Competition and social control in science: an essay in theory construction." *Sociology of Education* 41 (1968): 123-40.

Costello, Dennis, and Paul Rappaport. "The technological and economic development of photovoltaics." *Annual Review of Energy* 5 (1980): 335-56.

Cotton, Margaret D. Solar Energy Research Institute. "The photovoltaic R&D program at SERI: an overview report." SERI/SP-612-965. January 1981.

Crawford, Susan. "Informal communication among scientists in sleep research." *Journal of the American Society for Information Science* 22 (1971): 301-10.

Daele, Wolfgang van den, Wolfgang Krohn, and Peter Weingart. "The political direction of scientific development." In *The social production of scientific knowledge,* edited by Everett Mendelsohn, Peter Weingart, and Richard Whitley, pp. 219-42. Dordrecht, Holland: D. Reidel Pub. Co., 1977.

Denman, Scott, and Ken Bossong. "Corporate takeover of solar energy." *Business and Society Review* 34 (1980): 47-52.

DiMaggio, Paul, and Walter Powell. "Institutional isomorphism and collective rationality in organizations." *American Sociological Review* 48 (1983): 147-60.

Dietz, Thomas, and James Hawley. "The impact of market structure and economic concentration on the diffusion of alternative technologies: the photovoltaics case." In *Social constraints on energy--policy implementation,* edited by Max Nieman and Barbara J. Burt. Lexington, Mass.: Lexington Books, 1983.

Eckhart, M. T. "Assessment of solar photovoltaic industry, markets and technologies." Springfield, Va.: National Technical Information Service, 1978.

Ethridge, Mark. "The U.S. solar industries and the role of petroleum firms." American Petroleum Institute, research study #018, 1980.

Evan, William. "The organization set: toward a theory of interorganizational relations." In *Approaches to organizational design,* edited by James Thompson, pp. 173-88. Pittsburgh: Pittsburgh University Press, 1966.

Fallows, Susan. "The nuclear waste disposal controversy." In *Controversy: politics of technical decisions,* edited by Dorothy Nelkin, pp. 87-110. Beverly Hills: Sage, 1979.

Fan, John C.C. "Solar cells: plugging into the sun." *Technology Review,* August/September 1978, pp. 14-35.

Flavin, Christopher. *Electricity from sunlight: the future of photovoltaics.* Worldwatch Paper 52: December 1982.

Ford Foundation. *Nuclear power: issues and choices.* Nuclear Energy Policy Study Group. Cambridge: Ballinger, 1977.

Galbraith, John K. *The new industrial state.* New York: New American Library, 1967.

Ganz, Carole. "Linkages between knowledge creation, diffusion, and utilization." *Knowledge: Creation, Diffusion, and Utilization* 1 (1980): 591-612.

Gera, Ferruccio. "Geochemical behavior of long-lived radioactive wastes." ORNL-TM-4481. July 1975.

Gonzales, Serge. "Host rocks for radioactive waste disposal." *American Scientist 80 (1982):* 191-200.

Greenwood, Ted. "Radioactive waste management in the U.S." Aspen Institute Conference on Resolving the Social, Political, and Institutional Conflicts over the Permanent Siting of Radioactive Wastes. (Cambridge, Mass.: November 1979.)

Groves, Leslie. *Now it can be told: the story of the Manhattan Project.* New York: Harper and Brothers, 1962.

Hagstrom, Warren. *The scientific community.* Carbondale: Southern Illinois University Press, 1965.

Halloman, J. Herbert. "Government and the innovation process." *Technology Review* 81 (1979): 30-41.

Hannay, Bruce, and Robert McGinn. "The anatomy of modern technology: prolegomenon to an improved public policy for the social management of technology." *Daedalus* 109 (1980): 25-53.

Hench Report. "The evaluation and review of alternative waste forms for immobilization of high-level radioactive wastes." 20 August 1979.

Herwig, Lloyd O. "Report on photovoltaic research and technology in the U.S." In *Proceedings of the International Conference on Photovoltaic Power Generation* (Hamburg, Germany, 1974), edited by Helmut Loesch, pp. 29-42.

Hewlett, Richard. "Federal policy for disposal of radioactive wastes from commercial nuclear power plants." U.S. Department of Energy, unpublished paper, 1979.

Horwitch, Mel. "Designing and managing large-scale public-private technological enterprises: a state of the art review." *Technology in Society* 1 (1979): 179-92.

Interagency Review Group on Nuclear Waste Management. "Subgroup report on alternative technology strategies for the isolation of nuclear waste." U.S. Government Report TID-28818 (draft). October 1978.

_____. "Report to the president by the Interagency Review Group on Nuclear Waste Management." U.S. government report TID-29442. March 1979.

Jagtenberg, Tom. "Mission orientation in science." Unpublished master's thesis, Manchester University, 1975.

Johnston, Ron, and Dave Robbins. "The development of specialties in industrialized science." *Sociological Review* 25 (1977): 87-108.

Kamien, Morton, and Nancy Schwartz. "Market structure and innovation: a survey." *Journal of Economic Literature* 13 (1975): 1-37.

Kasperson, Roger. "The dark side of the radioactive waste problem." In *Progress in resource planning and environmental management,* edited by T. O'Riordan and K. Turner, pp. 133-63. Vol. 2. New York: Wiley, 1980.

Kast, Fremont, and James Rosenzweig. "Organization and management of space programs." In *Advances in space science and technology*, edited by F. Ordway, pp. 273-364. New York: Academic Press, 1965.

Kelly, Henry. "Photovoltaic power systems: tour through alternatives." *Science*, 10 February 1978, pp. 634-43.

Kerr, Richard A. "Nuclear waste disposal: alternatives to solidification in glass proposed." *Science*, 20 April 1979a, pp. 289-91.

_____. "Geologic disposal of nuclear waste: salt's lead is challenged." *Science*, 11 May 1979b, pp. 603-6.

Kevles, Daniel. *The physicists*. New York: Random House, 1971.

Klingsberg, Cyrus, and James Duguid. Assistant Secretary for Nuclear Energy. "Status of technology for isolating high-level radioactive wastes in geologic repositories." DOE/TIC 11207 (draft). October 1980.

_____. "Isolating radioactive wastes." *American Scientist* 80 (1982): 182-90.

Lamont, Lansing. *Day of trinity*. New York: Atheneum, 1965.

LaPorte, Todd. "Managing nuclear waste." *Society*, July/August 1981, pp. 57-65.

Lazarsfeld, Paul, Bernard Berelson, and Hazel Gaudet. *The people's choice*. New York: Columbia University Press, 1948.

Lenski, Gerhard, and Jean Lenski. *Human societies*. New York: McGraw-Hill, 1974.

Leong, John, and Satyen Deb. "Advances in the SERI/DOE program on CdS/Cu_2S and CdS/Cu-ternary photovoltaic cells." Fifteenth IEEE Photovoltaic Specialists Conference (Kissimmee, Florida, May 1981). DE81025858. New York: IEEE, 1981.

Lindmayer, Joseph. Personal communication. 1980.

Lipschutz, Ronnie. *The problem of radioactive waste*. Cambridge: Ballinger Pub. Co., 1980.

Loferski, Joseph. Personal communication. 1980.

Madrique, Modesto. "Solar American." In *Energy future*, edited by Robert Stobaugh and Daniel Yergin, pp. 183-215. New York: Random House, 1979.

Mansfield, Edwin. "Contribution of R&D to economic growth in the U.S." *Science* 175 (1971): 477-86.

Marcson, Simon. "Research settings." In *The social contexts of research,* edited by Saad Nagi and Ronald Corwin, pp. 161-92. New York: Wiley, 1972.

Martindale, Don. *The nature and types of sociological theory.* Boston: Houghton Mifflin, 1960.

Maycock, Paul, and Edward Stirewalt. Photovoltaics: sunlight to electricity in one step. Andover, Mass.: Brick House Pub. Co., 1981.

Mayr, Otto. "The science-technology relationship as a historiographic problem." *Technology and Culture* 17 (1979): 663-73.

Mazur, Allan, and Elma Boyko. "Large-scale ocean research projects: what makes them succeed or fail." *Social Studies of Science* 11 (1981): 425-50.

McCarthy, Gregory, et al. "Interactions between nuclear waste and surrounding rock." *Nature,* 18 May 1978, p. 216.

Merton, Robert. *The sociology of science.* Chicago: University of Chicago Press, 1973.

Morris, David. "The dawning of solar cells." Institute for Local Self-Reliance, 1975.

Mulkay, Michael. "Sociology of the scientific research community." In *Science, technology, and society,* edited by Ina Spiegel-Roesing and Derek de Solla Price, pp. 93-148. Beverly Hills: Sage, 1977.

_____. "Sociology of science in the West." *Current Sociology* 28 (1980): 1-117.

Nealey, Stanley, and John Hebert. "Public attitudes toward radioactive wastes." In *Too hot to handle? Social and policy issues in the management of radioactive wastes,* edited by Charles A. Walker, Leroy C. Gould, and Edward J. Woodhouse, pp. 94-111. New Haven: Yale University Press, 1983.

Nelson, Richard, and Sidney Winter. "In search of useful theory of innovation." *Research Policy* 6 (1977): 36-76.

Noble, David. *America by design.* New York: Oxford University Press, 1977.

Norman, Colin. *Knowledge and power: the global research and development budget.* Worldwatch Paper 31: July 1979.

O'Connor, James. *The fiscal crisis of the state.* New York: St. Martins, 1973.

Office of Nuclear Waste Isolation. Columbus Program Office. Richland Operations Office. "National Waste Terminal Storage Program." ONWI Technical Program Plan. ONWI-19 (rev. 1). December 1979.

Office of Technology Assessment. "Energy conversion with photovoltaics." In *Application of solar technology to today's energy needs,* pp. 393-426. Washington, D.C.: U.S. Government Printing Office, 1978.

_____. *Government involvement in the innovation process.* Washington, D.C.: U.S. Government Printing Office, 1979.

_____. "National laboratories: oversight and legislative issues." (draft). 26 June 1980.

Olson, Mancur. *The logic of collective action.* Cambridge: Harvard University Press, 1965.

Paolillo, Joseph, and W.B. Brown. "How organizational factors affect R&D innovation." *Research Management* 21 (1978): 12-15.

Parsons, Talcott. "Suggestions for a sociological approach to the theory of organizations." *Administrative Science Quarterly* 1 (1956): 63-85.

Pelz, Donald, and Frank Andrews. *Scientists in organizations.* (Rev. ed.). Ann Arbor, Mich.: Institute for Social Research, 1976.

Perez-Albuerne, Evelio, and Yuan-Sheng Tyan. "Photovoltaic materials." *Science,* 23 May 1980, pp. 902-7.

Praeger, Denis, and Gilbert Omenn. "Research, innovation, and university-industry linkages." *Science* 207 (1980): 379-84.

Price, Derek de Solla. *Little science, big science.* New York: Columbia University Press, 1963.

Rapp, Friedrich. *Analytical philosophy of technology.* Boston: D. Reidel Pub. Co., 1981.

Ravetz, Jerome R. *Scientific knowledge and its social problems.* New York: Oxford University Press, 1971.

Redfield, David. "Photovoltaics--an overview." In *Solar energy in review,* edited by Frank von Hippel, 1980.

Reece, Ray. "The solar blackout: what happens when Exxon and DOE go sunbathing together?" *Mother Jones,* September/October 1980, pp. 28-37.

Robinson, George. *The Oak Ridge story.* Kingsport, Tenn.: Southern Publishers, 1950.
Rosenberg, Nathan. *Perspectives on technology.* London: Cambridge University Press, 1976.
Rothwell, R., and A.B. Robertson. "The role of communications in technological innovation." *Research Policy* 2 (1973): 204-25.
Rowden, Marcus. "Nuclear waste management: getting on with the job." In *Proceedings of the International Symposium on the Management of Wastes from the LWR Fuel Cycle* (Denver, July 1976), pp. 45-51.
Roy, Rustum. "Science underlying radioactive waste management: status and needs." In *Scientific basis for nuclear waste management,* edited by Gregory McCarthy, pp. 1-20. Vol. 1. Proceedings of a symposium of the Materials Research Society, 1979.

Sapolsky, Harvey. *The Polaris system development: bureaucratic and programmatic success in government.* Cambridge: Harvard University Press, 1972.
Sayles, Leonard, and Margaret Chandler. *Managing large systems: organizations for the future.* New York: Harper and Row, 1971.
Schon, Donald. *Technology and change.* New York: Delacorte Press, 1967.
Schumpeter, Joseph. *Capitalism, socialism, and democracy.* New York: Harper and Row, 1950.
Science. "Solar activist Denis Hayes heads SERI." 10 August 1979, pp. 563-64.
Seamans, Robert, and Frederick Ordway. "The Apollo tradition: an object lesson for the management of large-scale technological endeavors." *Interdisciplinary Science Reviews* 2 (1977): 270-304.
Shapiro, Fred C. *Radwaste.* New York: Random House, 1981.
Shrum, Wesley. "Scientific specialties and technical systems." *Social Studies of Science* 14 (1984): 63-92.
Smith, Jeffrey. "Photovoltaics." *Science,* 26 June 1981, pp. 1472-78.
Smits, Friedolf. "History of silicon photovoltaic cells." *IEEE Transactions on Electron Devices* ED-23 (1976): 640-43.
Solar Energy Research Institute. Annual report. Golden, Colorado, 1979.

_____. "Poli Si moves out of research, into development." *In Review: a SERI Monthly Update* 2 (1980): 3.

Stone, Jack, Ed Sabisky, Harv Mahan, Tom McMahan, and Frank Jeffrey. "The national photovoltaic program in amorphous materials." Fifteenth IEEE Photovoltaic Specialists Conference (Kissimmee, Florida, May 1981).

Studer, Kenneth, and Daryl Chubin. *The cancer mission: social contexts of biomedical research.* Beverly Hills: Sage, 1980.

Surek, T., A.P. Ariotedjo, G.C. Cheek, R.W. Hardy, J.B. Milstein, and Y.S. Tsuo. "Thin-film polycrystalline silicon solar cells: progress and problems." Fifteenth IEEE Photovoltaic Specialists Conference (Kissimmee, Florida, May 1981).

Sutton, John. "Organizational autonomy and professional norms in science: a case study of the Lawrence Livermore Laboratory." Paper presented at the Pacific Sociological Association, 1977.

Szalai, A. "Research on research and some problems of research bureaucracy." *Scientometrics* 3 (1979): 247-60.

Tornatzky, Louis, et al. "The process of technological innovation: reviewing the literature." National Science Foundation, 1983.

Tukey, John. *Exploratory data analysis.* Reading, Mass.: Addison-Wesley, 1977.

U.S. Department of Energy. Office of Energy Research. Office of Basic Energy Sciences. "BES Initiative FY 1980-89. Radioactive waste disposition." DOE/ER 0016. September 1979.

_____. Richland Operations Office. Columbus Program Office. "Report on geologic exploration activities." DOE-RL-C-14. January 1980a.

_____. Assistant Secretary for Nuclear Energy. Office of Nuclear Waste Management. "Nuclear waste management program summary document FY 1981." DOE/NE-0008. March 1980b.

_____. Office of Nuclear Waste Management. Final Environmental Impact Statement. "Management of commercially generated radioactive wastes." 3 vols. DOE/EIS-0046F. October 1980c.

_____. Assistant Secretary for Conservation and Renewable Energy. "Photovoltaic energy systems: program summary." DOE/CE0012. January 1981.

U.S. Energy Research and Development Administration. "Ceramic and glass radioactive waste forms," ERDA workshop (Germantown, Md., January 4-5, 1977).

U.S. Environmental Protection Agency. Ad hoc panel of earth scientists. "State of geologic knowledge regarding potential transport of high-level radioactive waste from deep continental repositories." EPA/520/4-78-004. June 1978.

U.S. Geological Survey. Circular 779. "Geologic disposal of high-level radioactive wastes--earth science perspectives." J.D. Bredehoeft, A.W. England, D.B. Stewart, N.J. Trask, and T.J. Winograd, 1978.

Utterback, James. "The process of technological innovation within the firm." *Academy of Management Journal,* March 1971, pp. 75-78.

Wagner, Sigurd. Personal communication. 1982.

Weick, Karl. "Educational organizations as loosely coupled systems." *Administrative Science Quarterly* 21 (1976): 1-19.

Williams, Jerome. "What's SERI?" In *Solar Energy Handbook,* pp. 68-70. Reprint. Washington, D.C.: Department of Energy, 1979.

Williams, J., and W. Stephenson. "Solar technology 1979--an overview." *Mineral Industries Bulletin* 22 (1979): 1-12.

Willrich, Mason, and Richard K. Lester. *Radioactive waste: management and regulation.* New York: Free Press, 1977.

Wilson, Carroll. "Nuclear energy: what went wrong?" *Bulletin of the Atomic Scientists* 55 (1979): 13-17.

Wirt, John, Arnold Lieberman, and Roger Levien. *R&D management: methods used by federal agencies.* Lexington, Mass.: Heath, 1975.

Wolf, Martin. "Historic development of photovoltaic power generation." In *Proceedings of the International Conference on Photovoltaic Power Generation* (Hamburg, Germany, 1974), edited by Helmut Loesch, pp. 45-65.

_____. "Historical development of solar cells." In *Solar cells,* edited by Charles Backus, pp. 38-42. New York: IEEE, 1976a.

_____. "Photovoltaic solar energy conversion." *Bulletin of the Atomic Scientists* 32 (1976b): 26-33.

Wuthnow, Robert. "The moral crisis in American capitalism." *Harvard Business Review,* March/April 1982, pp. 76-84.

Wuthnow, Robert, Wesley Shrum, and James Beniger. "Networks of scientific and technological information exchange in the technical innovation process." National Science Foundation. August 1982.

Ziman, John. "The collectivization of science." *Proceedings of the Royal Society of London* B 219 (1983): 1-19.

Index